T0167275

Telecom Expense Management for Large Organizations

A Practical Guide

Luiz Augusto de Carvalho
and
Claudio Basso

Version 2.0

iUniverse LLC
Bloomington

Telecom Expense Management for Large Organizations
A Practical Guide

iUniverse books may be ordered through booksellers or by contacting:

iUniverse LLC
1663 Liberty Drive
Bloomington, IN 47403
www.iuniverse.com
1-800-Authors (1-800-288-4677)

Because of the dynamic nature of the Internet, any web addresses or links contained in this book may have changed since publication and may no longer be valid. The views expressed in this work are solely those of the author and do not necessarily reflect the views of the publisher, and the publisher hereby disclaims any responsibility for them.

Any people depicted in stock imagery provided by Thinkstock are models, and such images are being used for illustrative purposes only.
Certain stock imagery © Thinkstock.

ISBN: 978-1-4917-2002-8 (sc)
ISBN: 978-1-4917-2001-1 (hc)
ISBN: 978-1-4917-2000-4 (e)

Library of Congress Control Number: 2014900269

Printed in the United States of America.

iUniverse rev. date: 01/22/2014

Contents

Preface

Our objective in writing this book is to provide the reader with a practical guide addressing the most common issues associated with telecom expense management (TEM) in large organizations. This book is aimed at the telecom manager, and the techniques described here are practical and easily applicable.

Note that from our perspective, telecom expense management encompasses much more than just managing telecom bills and contracts. The way we see the issue, telecom expense management has almost the same meaning as telecom cost management (TCM). There is a trend in the marketplace to limit the meaning of telecom expense management to managing bills and billing, but our view is that billing systems are just a part of the bigger process. Throughout this book, we are going to detail this view.

The book is divided into fifteen chapters, each of which deals with a specific aspect of managing costs in telecommunications infrastructures; it falls into five sections: managerial issues, managerial processes, bill and billing processing, traffic analysis, and specific aspects.

The first four chapters are concerned with the structural aspects of managing telecom in large organizations. Chapter 1 explains our view about how the telecom area should be organized to enable the best results. Chapter 2 discusses strategies for negotiating telecom services in order to achieve good prices and easy management. Chapter 3 discusses the policies and governance strategies that are basic to having a well-controlled telecom structure, and chapter 4 discusses the delicate balance between effort and benefits of the control.

1. Organizing the telecom area administratively
2. Sourcing and procurement
3. Policy and governance
4. Cost benefits of expense management methods

The next section is concerned with the main processes associated with managing the telecom structure. Chapter 5 discusses managing the inventory of assets and services, which in our view is the basis over which all control is developed; chapter 6 deals with the process of adding, removing, and changing the assets and services contracted. Chapter 7 encompasses the processes and strategies associated with managing contracts, and chapter 8 discusses the process of registering problems internally and externally (with the providers).

5. Asset and service inventory management
6. Service ordering and change control
7. Contract management
8. Help desk management

The third section discusses what is more commonly known as telecom expense management. Chapter 9 deals with the processing of the invoices, chapter 10 focuses on the billing auditing processes, and chapter 11 discusses how billing systems can become an effective control instrument.

9. Invoice processing
10. Auditing
11. Billing systems

The fourth section discussess how traffic should be controled and managed. In our view, managing traffic is the core of any telecom cost management strategy. Chapter 12 explains how traffic should be controlled and discusses the main strategies to reduce telecom costs.

12. Traffic analysis and optimization

The final section deals with three specific issues that, in our view, deserve to be discussed in more detail. Chapter 13 is concerned with mobile devices and applications, which is a growing issue in the telecom control effort. Chapter 14 addresses an organization's risk management (such as lack of connectivity), and finally, chapter 15 discusses how the information managed by the telecom area should be presented and distributed in order to be effective.

13. Mobile device and mobile application management
14. Risk management
15. Reports and analysis

This book is not meant to be read linearly like a story book; it is more like a manual, where you can jump directly to the chapter dealing with a particular issue. Of course, a linear reading is possible, given the fact that earlier chapters support or complement later ones. We hope you enjoy reading it; we also hope it will become a useful instrument for you. This area of expertise lacks a broad source of literature, and the lessons learned by the professionals in the field are rarely documented or shared. We made an honest effort to document and systematize this practical knowledge.

Chapter 1: Organizing the Telecom Area Administratively

Initially, we would like to focus on four administrative strategies that are usually associated with the effectiveness of a telecom area:

- voice and data controlled together
- telecom as part of the organization's information technology structure
- controlling telecom in a centralized way
- deployment of control systems and consequent processes

Of course, these are general guidelines, and specific contexts may impose different strategies; however, these four points usually are the way to go. Subsequent sections will discuss the reasons.

1.1 Voice and Data Controlled Together

Although this strategy may seem obvious, some organizations do not control voice and data in a unified way. In today's technological environment, it is very advisable to have a control structure wherein the same area within the organization manages both voice and data contracts. There are four main reasons for this:

1) There is a technological trend to unify voice and data services over the same equipment and protocols.
2) The providers of both services are usually the same.

1

3) The billing control processes are similar.
4) The technical and administrative skills required from the people involved in the control of both services are very similar.

These elements are at the root of the trend to unify the control of telecom resources (voice and data) under one area of an organization. That is by itself a very important administrative decision, which tends to improve the quality of resource management. Very often, by just unifying the control of voice and data resources, overall costs can be reduced.

Some organizations control mobile resources and devices separately; our view is that these devices should be controlled by the same group within the organization responsible for controlling data and voice resources.

1.2 Telecom as Part of the Organization's Information Technology Structure

The structure and responsibilities of managing telecom within a large organization have changed over time as shifts in technology took place. Generally, the objective has been to bring telecommunications into the IT sphere of influence, unifying voice and data disciplines in a telecommunications/network area.

The migration of the telecom area from general services to IT is an administrative trend that follows the technological trend of integrating voice and data over the same network.

In addition, managing a telecom infrastructure in today's large organization requires skills more often found in people working in the IT department than anywhere within the organization.

1.3   Controlling Telecom in a Centralized Way

Telecom management is one area where centralized control tends to generate better results. Of course it isn't an absolute truth; multinational companies must balance the benefits of centralized management against the difficulties of managing infrastructures in different countries with different languages, currencies, and cultures.

It is our experience that policies and standards should be defined globally as far as possible. This creates an environment where teamwork and cross-regional support are possible, greatly enhancing the efficiency of the human capital deployed across the organization.

Here we emphasize the need to have unified inventory databases, processes, and technological standards. Sometimes, several arms of a large organization spread around the world do not understand the benefits of unified policies and standards. Usually, the telecommunications team in each country tends to believe that its own ways are the best, but anyone who has managed a multinational telecommunications area knows that having standards is better, even if they are not going to be optimal in every environment. When telecom management is centralized, it leads to the following benefits:

- better prices (usually due to global negotiation, where the full weight of the organization is brought to the table, yielding better discounts)
- better control (when only one group is responsible for telecom resources, it usually reduces problems such as overcharges, overlaps, and having unidentified resources or resources that are not used)
- lower operational costs (when headcounts are reduced, there is a consequent reduction in personnel costs)

Centralizing control usually enables the organization to identify its telecom expenses. That fact alone is usually enough to justify centralization, because it shows how much telecom represents within the IT/infrastructure budget and keeps the subject on management's radar.

3

In more general terms, we have to keep in mind that telecom is a logistic system, and as such, the whole may be more than the sum of the parts.

It would be interesting to insert a caveat into the argument here that centralized management doesn't necessarily mean a centralized operation. If you have the right tools, you may be able to control and contract in a centralized way and yet keep the operation distributed, enabling different telecom teams to operate in different countries, for example.

This is feasible, as long as you manage to make all teams use the same management tools, under a defined hierarchical framework. That means that the local telecom teams may have some autonomy to contract telecom resources (the ones not covered for the worldwide contract, for example), but they have to include each contract and resource in a corporate telecom management tool in such a way that headquarters can see all the telecom expenditures and all resources contracted in all countries. The local teams will see only their own expenditures and resources.

Therefore, we may divide the term "centralization" into two types: financial and technical. Even if operational aspects force technical decentralization, financial centralization remains crucial. The centralized telecom management has to keep track of what is contracted and how much it is costing.

Financial centralization refers to the following:

- centralized resource inventory (including data, voice, and mobile resources)
- centralized contract inventory (including voice, data, mobile services, and maintenance)
- centralized telecom bills (even if received in different countries, all bills would be included in a common tool in a standardized framework, allowing centralized control)
- centralized billing system
- centralized bill auditing process (at least in a country basis)

Technical centralization refers to the following:

- centralized help desk for telecom issues
- centralized point of contact with the telecom providers
- centralized point of contact for equipment maintenance
- centralized network operational center (NOC)

1.4 The Deployment of Telecom Expenditure Control Software

It is important to understand that these concepts are applicable whether you have specific software or just use a spreadsheet to control your data. In this section, we will discuss the benefits of deploying software to control telecom expenditures.

Many companies control telecom costs manually, using spreadsheets; there are two problems with this strategy:

- The entities are interrelated, and in most cases, it is difficult and time consuming to control them. For example, a last mile is associated with an address, a contract, a cost center, a service provider, and equipment. Every time you change one of these entries in the spreadsheet, you may have to adjust the associated ones. When these adjustments must be done manually, they are often missed, and the database becomes outdated.
- Even when stored in a shared directory, it is difficult to access information (what information is where?) and to keep track of changes (which versions of the spreadsheets are the updated ones?).

In addition, using spreadsheets isn't helpful when linking the databases to the main groups of processes associated with managing a telecom area:

- control of the contracts and bills
- interacting with the providers (ordering new resources, canceling resources, changing resources, and maintenance)

- generating reports with financial association for several criteria (addresses, providers, contracts, business units, cost centers, last mile)

The lack of automatic linkage between these processes and the spreadsheets is a guarantee that the data depositories will soon become outdated.

Another important consideration that is very often overlooked is the fact that using a spreadsheet makes it difficult to know what information is where, which in turn creates the situation where an organization must rely on a specific individual to find specific data. The use of spreadsheets also makes it easier for a disgruntled employee to sabotage the organization's data.

Those are the reasons why using telecom expenditure control software (such as Northridge, Sentinel, Tangoe, or TRMS) can be very helpful. They support all the processes in regard to inventory control, bill processing, service ordering, and maintenance control. This is why these tools are usually worth having. Nevertheless, it is important to keep in mind that these tools helps with control, but they don't solve the problems by themselves.

Therefore, it is important that the people in charge of controlling telecom expenses understand that proper processes, contracting strategies, and trained personnel are just as important in controlling telecom costs as having the right software package. Tools for summarizing telecom expenses can be an important (and sometimes essential) part of controlling expenses, but they are still just one part of the process.

A common mistake when implementing this kind of tool is the expectation that it will solve all problems. The providers of the tools usually include an initial implementation cost, which covers populating the databases and training the users. This cost only includes data gathering and training; it doesn't cover processes or contractual adjustments with the telcos. Those issues are usually the organization's responsibility.

Only the organization holds the power to negotiate its own contracts and command its people to change the way they work. The providers of the tools may act as advisors (although, in most implementations, this isn't part of their mandate), but at the end of the day, the organization has the responsibility and the means to make the necessary adjustments.

Never assume that the problem can be completely outsourced just because you contract a tool. Even if you contract a consulting company to implement the tool, you have to be realistic about your expectations and understand that the consulting company will only gather the data, populate the databases, train the people, design how the processes are supposed to be, and determine how the contracts should be renegotiated. What a consulting company usually can't do (and isn't supposed to do) is to command the people to follow the processes and negotiate the contracts directly with the service providers. This is your responsibility (and it is not easily transferable).

Depending on the scenario, having a control tool may not be optional, if you manage several business units operating in several countries, each one with several telecom teams with autonomy to contract and cancel resources, having a tool may be the only way to balance the act between autonomy and centralized control.

Tools like that would give you accurate cost verification, at the same time allowing each business unit to keep its operational independence (able to contract and cancel the telco's services), as long as it complies with a standardized framework (provided by the software). A typical tool would cover the following items:

- physical control of the business unit's addresses, including number of people and workstations per location
- control of all the organization's data connections, guaranteeing that each connection (whether leased line, frame-relay, MPLS, etc.) is clearly identified (end A, end B, last mile, equipment, committed and extended information rates [CIRs and EIRs], nominal bandwidth, technology, IP, DLCI, service provider, contract and cost), and also guaranteeing that all information can be consolidated by address, business unit, country, and cost center
- control of all interaction with the service providers, automating all processes linked with requesting new services, requesting

changes in existing ones, requesting maintenance, and canceling services; that increases control, reduces mistakes, and standardizes and simplifies procedures
- control of the costs in segmented or consolidated ways

The tool should also address the following issues:

- give telecom country managers a complete view of the infrastructure under their responsibility, and at the same time, allow the headquarters managers to have a complete view of telecom costs for all countries, normalized by currency (an ideal tool should be multilevel, multicountry, and multicurrency)
- control the physical locations where the organization (and its business units) has the ends of its connections installed (normalizing the addresses avoids problems such as different business units located in the same physical address contracting duplicated connections)
- control the business units, associating them with their physical locations, and in this way control the business unit's points of presence (POPs)
- control the existing connections, identifying the end A, end B, bandwidth capacity (CIR, EIR, and nominal), cost, and contract of each connection (a connection encompasses equipment and last mile costs)
- control all processes of contracting, changing, maintaining, and canceling a connection, guaranteeing that the quality of service (QoS) provided is the one defined by the service level agreement (SLA)
- control the telco's invoices, guaranteeing that the values charged are the ones defined by the contracts
- control the values paid, regardless of the currency in which it was contracted

The tool should also control the following entities:

- addresses and Zip codes
- business units
- business unit points of presence

- connections (dedicated or not)
- contracts
- cost centers
- equipment
- invoices
- last miles
- service providers
- telco interactions (requests for maintenance, cancellations, provisioning, changes)

Chapter 2: Sourcing and Procurement

Contracting and negotiating services is part of the day-by-day work of any telecom area in a large organization. But we have to keep in mind that we have two types of negotiation processes: retail negotiation and wholesale negotiation. Retail negotiations with service providers are common. In these day-to-day negotiations, proposals and requests for individual resources are made regularly, and technical and economic aspects are discussed.

Wholesale negotiations, on the other hand, are much more challenging and usually encompass more resources (trunks, circuits, or other equipment). Negotiating a large contract encompasses strategic aspects not present in retail negotiations. These negotiations deal with aspects such as high values, a lengthy transition process, and quality of service issues; they are also carried out by more experienced telco negotiators. All these aspects make proper planning essential.

In this context, we must separate sourcing and procurement into two types: retail negotiations, which are associated with simple quotations and ordering, and wholesale negotiations, which in turn usually encompass complete renegotiation of a WAN contract. This chapter focuses on wholesale negotiations, including strategies and concepts.

Wholesale negotiations usually take place at intervals not smaller than three years. Even longer intervals often elapse between contract reviews; this is due to the habit of just renewing existing contracts instead of opening a completely new negotiation, including potential providers.

This "law of least effort" of choosing to stay in the same contract, even when it may not be the best choice, is one of the main reasons why we find so many telecom contracts with higher-than-expected prices. All large organizations have some level of inertia that has to be overcome to allow a proper wholesale negotiation. *Regular negotiations are key to achieve the best results; renewing an existing contract may be practical but rarely yields the best price.*

Considering the effort involved, IT or telecommunications managers who need to renegotiate a large telco contract should initiate the process at least one year before the actual renewal date. This is typically the time required to make a proper assessment of the current situation in terms of traffic per service and define the strategy, identify appropriate designs, and prepare a detailed request for proposal (RFP). Therefore, when dealing with service providers, it is absolutely crucial to have a very good and updated view of the traffic in terms of volume, interest, and profile. Spreadsheets like the one shown in the following picture have to be updated and understood by heart:

Type of call	Calls	Minutes	Value	% minutes	% Value
Collect calls	2,660	2,399.00	0.00 €	0.01%	0.00%
fix to mobile Intra state calls	135,220	220,551.00	168,161.59 €	0.90%	5.68%
fix to mobile Inter state calls	454,416	472,780.00	462,995.68 €	1.93%	15.65%
fix to mobile Local	314,205	486,268.00	369,043.78 €	1.98%	12.47%
Fix intra state	4,017,093	3,877,638.00	404,831.09 €	15.80%	13.68%
Fix inter state	11,752,302	10,003,687.00	998,390.48 €	40.77%	33.74%
Fix region	477,263	437,190.00	43,586.61 €	1.78%	1.47%
Fix - Local	5,667,798	5,190,523.00	488,815.22 €	21.15%	16.52%
International	3	16.00	51.00 €	0.00%	0.00%
Incoming calls local	708,539	3,690,752.00	0.00 €	15.04%	0.00%
Incoming calls long distance	41,322	156,267.00	23,117.41 €	0.64%	0.78%
tool free	543	974.00	0.00 €	0.00%	0.00%
TOTAL	**23,571,364**	**24,539,052.00**	**2,958,992.86 €**		

The negotiation process must take in consideration several key factors that are typically driven by the make-up and culture of the organization. For example, we must evaluate the flexibility and willingness of the organization in regard to absorbing new services and new functionalities.

From the above facts, it is clear that some sort of strategy or policy is required when executing a wholesale negotiation. The larger and more geographically dispersed the organization, the greater the need for a strategy. The definition of strategies and policies prior to negotiation should address at least the following points:

- a clear statement of the organization's global approach to service providers; here are a few examples:
- one global provider (or one provider for each country or region)
- providers with local capabilities and infrastructure in areas of operation
- the best use of local providers in their areas of operation
- local or universal language capability in regions of the globe
- finance capabilities, currency requirements, and payment terms
- provisioning and sales cycle time frames
- an aggregation of all communication spending up for renewal by category, with line items broken out for each of the following:
- outsource options
- voice, mobile, private, fixed, and public switched telephone network (PSTN)-based services
- conferencing services, including Web, voice, and video
- video streaming services
- remote access services
- data connectivity
- high availability
- a clear internal understanding and communication regarding network capacity and QoS requirements
- positioning the ideal topology based on flow analysis and geographical realities driven by political, organizational, and other regional influences
- positioning at least two transport strategies[1]
- defining specific contractual elements up front, including
- cancellation and penalty clauses, and
- accounts payable processes and dispute arrangements
- SLA components:
- penalties and contract cancellation conditions

---

[1]   This might have to be regionally based.

- technology refresh options
- help desk performance metrics
- customer premises equipment (CPE) ownership and management, which regional regulations might dictate
- a standard price list[2]

It is our experience that the people negotiating with service providers often do so without proper preparation; they occasionally have the mind-set that would be more adequate for a retail negotiation. It is also very common to find people negotiating telco contracts whose view is purely economic, without any deep understanding of traffic volume, interest, and profile. This lack of preparedness usually generates two very typical approaches:

- **The "kick the telco" approach:** There is no complication; we gather quotes from all providers and just choose the cheapest one.
- **The "minimalist" approach:** There is no need for a detailed specification; we just ask them for their price list.

These views are not entirely wrong, and they may work for a retail negotiation, but they ignore a fundamental factor of this process: When negotiating with service providers, large organizations not only compare the prices per service, they also compare the prices of the different transport strategies (deployment of points of presence, for example) and achieve a balance between tariffs and minimum committed volumes. None of these aspects are present in retail negotiations.

Here are two examples:

---

[2] Standard price lists are often desirable in decentralized organizations and have many advantages in the areas of cost appropriation and financial planning. This standardization is often only possible in regional settings; even though it might provide a simplified operational structure, it is almost inevitable that some operational entities will pay a price penalty due to price standardization (like the paradox between standardization and optimal performance). That said, the advantages to central planning, financial visibility, and overall simplicity in a large multinational organization is significant and usually worth the effort.

- **Comparing the prices against the cost of a private voice network:** The cost of the spoken minute has to be compared not only with prices from other service providers but also with the cost to transport them through the private voice network.
- **Mobile traffic:** The cost of the spoken minute fix-mobile must be compared with costs from fixed trunks providers and also with the alternative of providing dedicated mobile trunks (or using GSM gateways) to transport mobile traffic.

Therefore, it is crucial to accurately map the traffic, understanding perfectly from where and to where the traffic flows, the volumes, the quality requirements, and the available transport strategies.

We are not just comparing the costs per service but also the costs per transport strategy; this is a fundamental point and makes a difference when negotiating with telcos. Besides the possibility of comparing different transport strategies, the only other alternative is the direct comparison between the same services. In this situation, each telco usually knows the limits of the others. The chances for driving down costs are limited. (This is usually associated with volume.) There is very little room for discounts. In this situation, there is a big chance that the current provider, who already has its investment paid for, offers the best price.

Although large organizations may have several telecom providers, usually they concentrate their business in a few (between two and three providers are responsible for more than 80 percent of all telecom expenditures).

However, we should beware of exaggerating this concentration of business. It may trigger overconfidence by the providers and a lack of consideration of other alternatives available in the market. It is an important aspect to be considered; we often hear comments like, "We don't want to have several providers." There is a fine line here, and no one should see having fewer providers as an end in itself. It is only worthwhile if it guarantees better prices, more simplicity, and negotiating leverage.

The basic fact is that from the point of view of getting better prices, it is convenient to allow all providers to offer their services, even when they do not cover all your sites or offer good prices for all their services (here the routing capability shows its importance). Ideally, when quoting, the logic should be, "Provider: Quote your best price for the services you offer where you have coverage." Never put "having few providers" as a prerequisite when doing a quotation. This is an easy path to higher prices. Of course, having the ability to route traffic makes all difference in the effectiveness of this approach.

Here it is important to emphasize that allowing potential providers to present their best prices where they have coverage (no obligation to quote all services for all points of presence) implies that you may consider the possibility of having several providers. But it doesn't necessarily mean contracting in this way.

Eventually, a quotation can be done that allows providers to offer services only where they have coverage. Later, you may narrow down the options to the ones with wider coverage, using the cheapest prices of all proposals as negotiating references.

Although common, contracting only one service provider may not be advisable. Having at least two main providers may be better for commercial and technical reasons. Commercial reasons include competition and the reduction in cost of moving services between existing suppliers, as opposed to bringing in a completely new service provider. From a relationship point of view, it is always good when a service provider knows there is a concurrent competing contract to where the organization can easily turn for its requirements.

Technical reasons are based on security and risk mitigation. For example, it is advisable to use backup circuits contracted from a different provider than the one providing the primary circuits.

Of course, contracting a complete WAN, including the CPEs, with more than one provider may have its difficulties. A provider may have technical or commercial restrictions that prohibit other providers from connecting resources to its devices. This problem is rare though. In most cases, the quotations are done assuming that each provider will have its own CPEs, although this will vary between regions of the globe.

The management of multiple contracts may be seen as problematic, but considering the fact that a typical telecommunications area already manages several contracts (although concentrating services in a few of them), this argument seems to lack substance. If the internal people are not capable of absorbing this small amount of additional effort, something may be wrong with the way the organization manages its telecommunications contracts.

Other very common argument against segmenting quotations is the fact that such segmentation tends to reduce volume and consequently reduces the discounts offered. This line of thinking has several problems:

- First, and most obvious, is the fact that quoting with several providers doesn't eliminate the possibility that providers able to offer all services in all areas will do so if they are capable, eventually even giving alternative costs per volume contracted.
- Second, this line of thinking doesn't take in consideration the already mentioned fact that we have to compare not only service providers and services but also transport strategies (one transport strategy may only be applicable in some areas).
- And finally, and maybe the most problematic, is the simple fact that we may have potential providers whose prices are very good but limited to some specific geographical areas or specific types of services (mobile calls, for example). Depending on the percentage of our services belonging to these types and within these areas, even if we pay a lot more for the other services outside those areas (not covered areas), the savings could still be substantial. In addition, we always can route the calls, avoiding the expensive services and redirecting them to providers whose prices are more reasonable (once again, note the importance of being able to route traffic).

The quote should be structured in such a way that the service provider must present a defined price, clearly indicating the following(in case of voice services):

- cost per minute per type of call
- charging granularity
- monthly fee for trunk subscription (if any)

- minimum committed volume (if any)
- costs for other services (for example, installation)
- periodicity and duration of the contract (typically monthly and not above thirty-six months)

In the case of data services, the following elements should be clearly indicated: monthly fee for last mile, including EIR, CIR, and nominal bandwidth monthly fee for CPE, including cost per port if necessary minimum committed volume (if any) costs for other services (for example, installation and management) periodicity and duration of the contract (typically monthly and not above thirty-six months)

In addition to the above, the services contracted today may not reflect the organization's actual needs, which may have changed since the last negotiation assessment. This is the main problem with the minimalist approach. Therefore, when preparing to renegotiate the WAN, it is crucial to accurately map the traffic, understanding perfectly from where to where it flows, the volume, quality requirements, and available transport strategies. In other words, before discussing technologies or contacting the potential providers, the telecommunications manager has to understand very well the organization's current and future needs in terms of traffic volume, interests, and profile.

Eventually, different minimum committed volumes imply a different set of tariffs. If it is the case, it has to be clearly stated. The clear indication of all these items enables visibility and transparency when managing the contract and comparing costs.

2.1 Negotiation Strategy

Planning the negotiation strategy is key to successful negotiations (remember the Latin saying *"Amat Victoria Curam"*). The process must consider several key factors that are typically determined by the make-up and culture of the organization. As an example, the telecommunications manager must evaluate the organization's disposition to manage and trust its providers. When planning a WAN negotiation, the following factors should be evaluated regarding the organization itself and its needs, the environment where the organization operates, and the providers available:

- agility required from the provider companies[3]
- capabilities of the organization in different countries or regions of the globe degree of centralization or decentralization of the IT infrastructure and decision-making
- degree of telecommunications deregulation in countries of operation around the globe language barriers degree of local knowledge of telecommunications industry in other areas of the globe
- ability or desire to manage multiple vendors and finance systems with different currencies, languages, cultures, and accounts payable environments
- differences of control and capabilities in regions of the world over different technologies like voice, mobile voice, and data networking, which are often the result of an organization's historical growth and acquisition strategy
- management's view of outsourcing
- available services; frequently, competitive local exchange carriers (CLECs) and incumbent local exchange carriers (ILECs) in different countries own the entire range of services, from mobile telephony to providing leased line services, utilizing their own infrastructure

When providers offer an entire range of services, the telecommunications manager may be able to aggregate services, increasing the overall size of the financial pot they are competing for. This strategy often makes possible the achievement of better prices (although having the downside of concentrating too many services into few providers).

Finally, there are control issues, which are another political aspect that organizations tend to ignore. Analyzing the existing control structures and confronting them with the strategy of contracting telecom services is an

---

[3] Telecommunications companies are very process-oriented, which results in organizations that are not very dynamic in their mode of operation.

important exercise. The people defining the negotiating strategy should conduct this analysis.

2.2 Interconnection Costs (Tariffs)

Managing a telecom structure in a large organization requires an understanding of tariffs. You must know exactly how much each type of service costs for each service provider (active contracts). It is helpful to prepare and update a spreadsheet such as the following:

Traffic fix to fix:

Type	Telco 1		Telco 2		Telco 3		Telco 4	
	Cost per minute	Granularity sec	Cost per minute	Granularity sec	Cost per minute	Granularity sec	Cost per minute	Granularity sec
Local	USD 0.050	30+6+6	USD 0.027	30+6+6	USD 0.053	60+60	USD 0.04	60+60
Long distance POP	USD 0.051	30+6+6	USD 0.083	6+6+6	USD 0.077	30+6+6	USD 0.10	60+6+6
Long distance outside POP	USD 0.113	30+6+6	USD 0.083	6+6+6	USD 0.077	30+6+6	USD 0.10	60+6+6

Traffic fix to mobile:

Tipo	Telco 1		Telco 2		Telco 3		Telco 4		Telco 4	
	Cost per minute	Granularity sec	Cost per minute	Granularity sec	Cost per minute	Granularity sec	Cost per minute	Granularity sec	Cost per minute	Granularity sec
Mobile local	USD 0.506	30+6+6	USD 0.66	30+6+6	USD 1.12	30+6+6	USD 0.32	30+6+6	USD 0.32	30+6+6
Mobile inter and intra state	USD 0.506	30+6+6	USD 0.66	6+6+6	USD 1.12	30+6+6	USD 0.55	30+6+6	USD 0.43	30+6+6

2.3 Benchmarking

Telecom managers must be aware of the telecom costs in their markets. This awareness should be organized in a spreadsheet to allow a clear understanding of the costs that can be negotiated. The prices disclosed by other telecom managers, consultants, and proposals can be incorporated in the spreadsheet for comparison. Often, this market intelligence can be paid for—many companies specialize in scanning the market and mapping the prices of different services. If your company operates in only one country, you probably already have a good view of the prices there. However, it is still a good practice to document these prices and periodically share this information with other companies (even if only informally). If you feel like you are with wrong references, buying market intelligence may be a good idea.

This type of comparison doesn't pretend to be absolute; its main objective is to provide minimum and maximum prices in the markets where your company operates. Of course, you should be aware that the real prices have to be achieved through a hard bargaining process. The references should be seen as achievable targets.

The following spreadsheet shows a comparison between the current prices of a given organization and the potential prices in a given marketplace (minimum and maximum prices).

Type	Tariff with taxes	Minutes (x1.000)	Current cost (USD)	%	Values of reference		Range of values	
					minimum (USD)	maximum (USD)	minimum (USD)	maximum (USD)
Outbound local								
Fix-Fix	0.0615	213	13,094.40	4.62%	0.0252	0.11	5,375.15	23,435.05
Fix-mobile	0.5532	19.7	10,917.60	3.85%	0.51	0.65	10,065.67	12,828.79
Outbound long distance								
Fix-Fix	0.0701	459.8	32,219.97	11.37%	0.0252	0.11	11,600.22	50,575.69
Fix-mobile	0.6307	33.8	21,327.35	7.53%	0.51	0.65	17,246.01	21,980.21
800								
Fix-Fix	0.0561	277.5	15,557.44	5.49%	0.0252	0.11	7,001.47	30,525.65
Fix-mobile	0.0561	66.8	3,743.34	1.32%	0.0252	0.11	1,684.65	7,344.90
Total			96,860.11	34.19%			52,973.17	146,690.28

As can be seen, this spreadsheet allows the quick calculation of the total cost to the organization for a given value. In this particular example, the current cost is USD 96,860.00. The spreadsheet shows that the minimum and maximum values practiced in this market are between USD 52,973.00 and USD 146,690.00. That means that the current prices are slightly below the average (USD 99,831.00). Therefore, it is clear that there is some potential for savings through tariff renegotiation. If the organization manages to negotiate the minimum tariffs available in this market, it could reduce its current costs by 45 percent (the difference between USD 96,860.11 and USD 52,973.00). Telecom managers should have this kind of general understanding, even if they can't do anything to change the scenario right away.

With all these ideas in mind, we can now structure the quotation process. The quotation should be structured in such a way that the service provider presents a defined price per connection, and these prices must be associated with defined parameters. This enables visibility and transparency when contracting new services or changing (or canceling) existing ones during the contract life cycle, which will surely occur.

2.4   Pitfalls to Avoid

When preparing an RFP or a request for information (RFI) for telecom services, it is very important to avoid some basic pitfalls. Here are some of the more common ones:

- **Price visibility:** Very often, service providers are reluctant to offer a price list, using the argument that if they present a defined price list, they will be punishing the client if the prices go down, because they will not be able to offer those prices. We don't buy this argument. The difficulties associated with visibility of the telecommunications costs of new sites, including the possibility of the prices going up instead down and the difficulty of verifying costs associated with a specific provider, clearly surpass the hypothetical benefits to the service provider of not disclosing prices. Therefore, it is absolutely crucial to clearly state the price of each service.

- **Paying for the gateways:** As a common marketing practice, vendors often prepare quotations of packet networks based only on the equipment and last mile costs. In this quotation strategy, the gateway cost has to be paid separately. This approach makes it difficult to clearly identify the connection cost per site and very often generates problems of gateway over-/undercapacity. The ideal is to avoid it when possible. However, decision is not always in the telecommunications manager's hands. If potential providers refuse to charge on a per connection basis, the telecommunications manager may not have a choice. We recommend always trying to do this. You may not succeed in 100 percent of the cases. Knowing the ideal does not mean you always will be able to do it.

- **Backup structure:** When contracting backup circuits, avoid contracting the main and backup circuits from the same provider. A single provider, even if the provider has two different physical infrastructures, can expose the organization to administrative problems that may affect the provider, such as strikes, bankruptcy, and so forth.

2.4.1   Paying for the Gateways

In packet networks, there is gateway sharing. In this context, the gateway is the last mile installed in the organization's main hubs in typical star networks, to where and from where the connections flow. In these scenarios, the connections share the gateways. It is very important that the capacity of the last mile does not exceed the capacity of the gateway (even though some statistical maneuvers allow that correlation to be different than 1:1); if they do, the flows will be limited to the capacity of the gateways. The nominal capacity of the last mile is valid only to the border of the telco cloud. In this situation, the organization will be paying for bandwidth in the last mile that it cannot use. A practical tip: never allow the gateway bandwidth to have a ratio with the summation of last mile bandwidth below 1:0.6. In other words, a gateway should not have a capacity below 60 percent of the summation of the associated last mile.

Frequently, the service provider implies that increasing the bandwidth of the last mile of one specific site will increase the available throughput. This is not always true. Gateway sharing limits this seemingly straightforward increase. It is important to be very aware of this likely source of wasted expense.

This situation is very common. An organization sometimes pays to increase the bandwidth (end A) without any significant improvement in the QoS. Very often, the gateway is oversubscribed, and the increase of the bandwidth in one end merely increases the existing rate of oversubscription. The organization pays more for no additional benefit.

To avoid this problem, the telecommunications manager should make sure that the summation of the CIRs contracted in the accesses cannot exceed the bandwidth available in the gateways, beyond 40 percent (unless it is an

Internet access and part of the flow does not go to the gateways). It is just a matter of doing the math or monitoring the actual usage.

This sharing of the gateway, along with the fact that the cost appropriation is usually on an equipment and last mile basis, makes it difficult to have a clear understanding of the cost associated with providing connectivity to each site. Using the framework of cost appropriation helps to clarify the specific parts of the cost breakdown. The costs should be broken down as follows:

- equipment and port A: usually charged having the cost associated with site A
- last mile A: usually charged having the cost associated with site A
- cloud: usually has a cost appropriation equaling zero
- last mile B: usually charged having the cost associated with site B
- equipment and port B: usually charged having the cost associated with site B
- connection and traffic: usually charged having the cost associated with the bandwidth of the PVC—(Private Virtual Connection) defined to site A

Because more than one connection (gateway sharing) uses the equipment and last mile B (gateway), it becomes difficult to accurately identify the total cost to provide connectivity to a specific site. It becomes necessary to proportionally divide the cost of the gateway among the sites using it.

Considering the importance of clearly identifying the cost to provide connectivity for each point of presence, when possible, the organization must require the service providers to present their prices end-to-end (connection basis). In a price format like that, all prices may be allocated to end A. For the shared last miles (gateways), the cost would be zero.

The best way to avoid the common problems caused by gateway sharing and the difficulty matching the pricing model to the cost factors just described is to contract the service in an entire connection basis (end-to-end). In this situation, the provider becomes responsible for guaranteeing throughput end-to-end, so it becomes its responsibility to adjust the gateways as connections are increased, reduced, or added.

The objective of a WAN is to provide throughput between two points. It is important to identify how much it costs to provide a given throughput between two specific sites. The cost to be analyzed is that of achieving the main objective of connectivity, which always requires complete connections (two points).

Although some cost structures consider end-to-end connection costs, the more common practice in data networks pricing is to only consider the access cost (last mile and equipment), charging the gateways separately, as if they were a normal last mile. Here are some typical types of services and their charging structures:

- **Leased lines:** The charging structure typically follows a function correlating distance and bandwidth. The cost usually covers both ends.
- **Optical fiber connection:** The cost is usually represented with a link between the distance and number of derivations (end-to-end).
- **Satellite connections:** The cost of this service usually correlates the volume in/out and antenna cost and a component associated with the hub cost (access and hub).
- **Dial-up access:** This normally follows a widely known price spreadsheet, which correlates destination (and distance), duration of the call, and sometimes the time of the day (or day of the week).

2.5   Practical Tips for Outsourcing Networks

When dealing with the subject of outsourcing WANs, *outsourcing* means contracting hardware and services outside the organization. It is important to understand this basic concept: Unless the organization owns all transmission means and equipment, and manages its own NOC to some extent, it will be outsourcing part of its network to a service provider (or many providers). Therefore, in the absolute majority of the cases, WANs are outsourced, even if only partially.

In the telecommunications context, *outsourcing* is used only when equipment and its management are contracted with the service provider. People do not usually say the network is outsourced when only the means (circuits) belong to a provider and the equipment belongs to the organization. From our perspective, this is a mistaken view. All of them are outsourcing processes, with the only difference being the degree.

It is very rare to find an organization that owns all WAN components and operates them itself. This kind of situation may happen in petroleum companies, railways, highway operators, and governments. These organizations usually build their own networks. Building their own network is typical for service providers (and even they may subcontract parts of their infrastructure).

In a more common situation, where the organization does not own its whole network, a WAN is outsourced on many different levels. It may have its transmission means contracted from a telco (which is not usually perceived as outsourcing) or its equipment rented, with maintenance contracted from a telco or hardware vendor. The organization may have its NOC services contracted with a telco or service provider. Finally, it can have its own telecommunications personnel contracted through a personnel outsourcer.

In most cases, we see a combination of these outsourcing scenarios. Although outsourcing has been in and out of fashion several times (and was sometimes even sold as a panacea), experience has shown we should have a more cautious approach to the pros and cons of this process.

It is important to clarify that there is no point in being for or against outsourcing. As mentioned, practically all organizations, to some extent, outsource their WANs. In the end, the point is to evaluate how each outsourcing strategy is to be implemented and to see the advantages and disadvantages of each particular strategy as it applies to your organization's requirements and culture.

The outsourcing of the means is practically the rule among organizations. Normally, only the companies that own tracks of land interconnecting their sites will build their own infrastructure, for example, petroleum and gas companies, railways, and highway operators.

Other exceptions where outsourcing is not considered a valid alternative are government organizations, whose considerations are not purely economic, such as to prevent reliance on public networks. This includes armies, security agencies, and natural disaster agencies.

Giving an additional emphasis to the analogy with the transportation system, we would have a situation where a company could contract a truck or railway company directly or contract a company like UPS or FedEx to manage its storehouses and transport their goods. In the same way that happens between ILECs/CLECs and carriers, a company like FedEx may subcontract a truck company, which, in turn, would use a public highway.

Outsourcing the equipment is also very common, but the way it is done may vary:

- rent the equipment and contract the maintenance and management from a hardware vendor
- contract the equipment and management from a telco
- rent the equipment and maintenance from a hardware vendor and the management from a service provider
- buy the equipment and contract the maintenance from a hardware vendor and the management from a service provider
- buy the equipment, contract the maintenance from a hardware vendor, and perform the management internally

Other combinations are possible, but these are the more common ones. At this point, we have to understand the conflicts of interest that can make some of these combinations unsuitable.

Initially, it is important to understand that equipment like switches, routers, or PBXs are essentially resources that enable the management of the telecommunication means. This management, in theory, is performed by selecting lowest cost routes and managing traffic flow. Because of that, there tends to be a conflict of interest between who provides the means (circuits and trunks) and who provides the equipment. We can illustrate this situation with two examples:

- **CENTREX service (PBX provided by the telco):** What interest would Telco 1 (the owner of the PBX) have to configure the PBX to redirect calls to Telco 2, when the call through Telco 2 has a lower cost? Assume (optimistically) that Telco 1 would allow the trunks belonging to Telco 2 to be installed in its PBX.
- **Bandwidth optimization resources:** What interest or diligence would a telco have to implement bandwidth optimization resources in its equipment if, when doing so, these resources will reduce the need for bandwidth and reduce the values paid by the client?

These examples highlight the issue of contracting equipment and means from the telco. When an organization adopts this strategy, it is quite likely that the telco that owns the equipment will deny access to other service providers. Even if it does allow other providers to install their trunks, it will resist requests to configure them to redirect traffic to other provider trunks (CENTREX service example). An organization's telecommunications manager may try to force them to play nicely together (very large organizations may be able to do that), or it may buy or rent its own equipment and link all providers to them. The equipment may or may not belong to the company and may or may not be managed directly by it. Some intermediary alternatives can be negotiated as well, such as you can see the configuration but not change it, or you can see and change it, but not beyond a point.

The outsourcing of network management is even more delicate in terms of conflict of interest. If the hardware provider is responsible for the hardware maintenance and network management, it will tend to be condescending with equipment problems (its own fault). In this scenario, the hardware provider will have a vested interest in disputes with the telcos and tend to attribute the problems to them, avoiding the fact that the problems were related to its own equipment. On the other hand, if the entity responsible for management is also responsible for the telecommunications means, the same problem described previously will occur, but with the issues reversed. We often hear the magical solution: Let's put the means, equipment, and management under the responsibility of the same entity. This is the worst-case scenario. Once we put the responsibility for identifying problems entirely in the hands of the service provider, it creates a situation where a service provider punishes itself financially when it identifies its own faults. This is definitely not an effective strategy for the organization. Surprisingly, this is a very common situation, and it tends to generate a high level of stress between the provider and the organization.

From our perspective, the ideal scenario is when the organization outsources parts of the structure to different providers, with one or more entities responsible for the means or circuits, one responsible for the equipment maintenance, and one for the network management.

A rarer situation is when the telecommunications department is itself outsourced. In this scenario, internal personnel are contracted through another company. This kind of situation can cause conflicts of interests. The ideal is never to contract internal people from a company that provides other services (network management, maintenance, or means).

The economic argument for outsourcing makes sense theoretically. An organization that provides a large number of telecommunications resources tends to have a smaller cost per served unit than one with a smaller volume. However, there are two problems with this logic:

- Very often, the companies that present themselves as outsourcers are not the owners of the network. They subcontract the infrastructure from one or more telcos. Although it is true that these companies tend to get smaller prices from the telcos due to volume, that does not necessarily result in better prices for the final clients. Very often, the operational cost of the outsourcer accounts for the difference between the value that the client would pay if contracting directly from a telco and the cheaper price that the telco originally gave to the outsourcer. This problem becomes more acute as the size of the organization grows, because the difference between what the organization can obtain negotiating directly with a telco and the value obtained by the outsourcer narrows significantly.
- If the outsourcer is a telco and actually owns the infrastructure, it will tend to treat the process as if it were a conventional proposal (selling circuits or means), just adding additional services to

manage the network by an in-house team. This tends to eliminate scale gains and just add services to the telco bill.

A company trying to decide between purchasing equipment and relying on a service provider must consider a combination of factors. The obvious comparison is between up-front capital and ongoing maintenance costs associated with purchasing equipment versus leasing similar functions from a service provider. Additionally, there are more intangible issues related to things like control; for example, level of responsiveness, bundling, and leverage (does the person who needs to fix the problem actually work for me, or am I dealing with a contractor over whom I have less leverage?). Generally, three interrelated factors qualify the feasibility of outsourcing a telecommunications structure:

- **Size of the company:** Generally speaking, for large corporations, it usually makes sense to buy equipment and hire staff to maintain it.
- **Rate of growth:** If a company is growing quickly, there is a danger of outgrowing the infrastructure before it is time to replace it. Outsourcing gives the company some flexibility when expanding.
- **Expertise of staff:** In terms of staff, some companies have a large telecommunications/IT department and like to do things in-house. Others prefer to stick to their core business and leave the care and feeding of their telecommunications networks to others.

There are some gray areas when deciding whether to outsource a telecommunication network or not. But we should keep in mind that the main motive for outsourcing is usually savings. However, when evaluating the potential savings, we should consider not only how much the organization is spending today, but how much it could be expending if doing everything in-house in the best way possible.

Most outsourcing evaluations are based on the actual company's expenditures compared with the foreseen outsourced price. The problem with this practice is the fact that, in most cases, the company could be doing a much better job in-house than it is actually doing, so the comparison between actual expenditures and outsourcing is not fair or in the company's best interest.

Therefore, in order to rationalize the decision of whether to outsource a network or not, it is absolutely crucial to identify the optimized in-house cost and then use this value to evaluate outsource alternatives. Using the actual cost as a reference can give you the wrong perspective about the benefits of the strategy.

Some may argue that savings would be achieved anyway, whether you use the current value as a reference or identify the possible minimal in-house cost. We particularly believe that having a correct base of reference for cost is crucial in this process. Only when the theoretical ideal cost is known can the real benefits of outsourcing be defined. Only then can you balance the internal effort of achieving the ideal costs without outsourcing and the outsource cost presented.

Very commonly, potential outsourcers do the homework of analyzing the WAN, identifying the potential gains, and defining the price based on the current cost. In this situation, the outsourcer will take advantage of the organization's lack of knowledge of its own needs and possibilities.

In summary, the decision to outsource encompasses many considerations. Knowing what is already outsourced, being aware of potential conflicts of interest, and having a thorough understanding of economic considerations are all important. It is only possible to fairly evaluate the cost benefits of various outsourced solutions when you have an accurate view of the costs to build and properly manage an in-house optimized infrastructure.

2.6   Managing Service Providers and Vendor Relationships

Dealing with service providers is an integral part of the telecom manager's job. Here it is important to emphasize some points that, although a bit obvious, are sometimes overlooked:

- You have to keep in mind that *you* are the person responsible for identifying your organization's needs; this isn't the service provider's job.
- Service providers aren't your friends; they represent a set of interests that may or may not be in line with your organization's interests (most often, they are not).
- Service providers usually manage to set an undue influence over the telecom manager; this is achieved through excessive interactions. A much better understanding of the market context may be obtained through participating in user groups and attending trade fairs.
- After all the conversation about partnerships among companies, at the end of the day, the job of service provider account managers is to get more money for their company's services, and your job is to get more services for less money.
- Any dealings with service providers should be strictly professional; situations where you or anybody of your team can be drawn into private conversations outside the workplace are highly undesirable.

Chapter 3: Policies and Governance

Having good operational policies and good governance is crucial to controlling telecom costs. The policies are the base over which the processes are structured, and governance encompasses the mechanisms the organization uses to ensure that its constituents follow its established policies.

A proper governance strategy is what guarantees compliance with agreed-to processes and provides for corrective action in cases where the rules have been broken, ignored, or misconstrued.

Therefore, we have to identify certain definitions that need to be understood:

1. Define the operational policy (basic rules and strategies).
2. Based on the policy, define the processes to be followed.
3. Define the tasks and responsibilities. Once the policies and processes are defined, implement the mechanisms to make sure that those policies and processes are followed.

Here, it is worth giving a practical example of what we mean. For example, here is the sequence if we want to define the process of receiving and processing bills:

Step 1: Define the operational policy.

Example of policies:

- All bills will be received via electronic media and downloaded directly from the provider website.

- Bills not make available on the provider website a minimum ten days before the payment day will not be paid on time, due to the fact that there isn't enough time for auditing.
- If the value charged is different from the value identified as correct by the auditing process by more than 7 percent, the organization will pay the value identified as the correct one and dispute the discrepancy with the provider.

Once we define the macro rules, we can define the processes.

Step 2: Define the processes to be followed.

Example process:

- The bill will be downloaded from the provider's website and stored in a defined directory where the auditing team will execute the recalculation of all calls.

Step 3: Define the tasks and responsibilities.

Break down the processes into small tasks, identifying who is responsible for doing what and when.

Once we have a clear definition of what needs to be done, how it should be done, and who is responsible for doing it and when, it becomes possible to enforce the rules, checking exactly what went wrong and why (governance).

3.1   Definition of Policies

The proper definition of operational policies is a prerequisite to the effective operation of a telecommunications infrastructure. These policies need to be clearly defined. The definition of these policies implies a very clear understanding of the organization's overall IT strategy. It is not our intention to make recommendations of specific policies, because those are dependent on each organization's specific needs. As mentioned, policies precede processes when organizing the operation of a telecom area. We are only listing some typical decision points:

- contracting strategy for equipment and means (same providers for equipment and means, or not)
- NOC operation (in-house or outsourced)
- resource acquisition (rent, buy, or lease/periodic hardware renew)
- support and maintenance strategy (permanent technical assistance contracts or as-needed requests)
- strategy to interact with users' requests via a help desk
- expense approval policy
- technological standards
- level of participation of the telecommunications area in the organization's IT decision process
- definition of telecommunications resource demand standards
- definition of the policy of resource control and documentation
- training policies
- global infrastructure standards
- regional policy decision points
- global policies affecting regulatory requirements
- global change control policy (usually part of an overall IT change control policy)
- internal cost appropriation
- policies for processing bills
- policies for bill discrepancy dispute

The definition of policies isn't a unilateral process; very few of them are exclusively within the IT/telecom. Most policies have to be discussed with providers or with internal users. Most often, you will need to interact with your service providers and internal clients to be able to set up the operational rules (policies).

Example policies follow:

- You can't define the bill discrepancy dispute policy without making your providers agree with it.

- By the same token, you can't unilaterally define the user profiles and the level of services made available to them; this kind of policy has to be discussed with the users.

Once defined, these policies should be properly followed. The organization must avoid changing definitions by default following certain circumstances or for the convenience of the moment. Ideally, these definitions should reflect a corporate posture and must be completely compliant with the overall IT policies and strategies. Of course, these policies should not be static. They should follow the changes in the organization's needs.

3.2 Definition of Processes

The processes required to manage the telecommunications infrastructure must be carefully planned. But planning is not enough. The processes need to be executed correctly and consistently. We can divide the processes of a telecom area into four main groups:

- **operational and technical:** keeping the structure running
- **cost control:** keeping track of how much the organization spends on telecommunications
- **documentation and administrative:** keeping track of resources and configurations
- **planning and expansion:** preparing for new services and matching demand with availability

These four groups may be subdivided as follows:

- operational and technical
- help desk operations
- voice trunks and data links monitoring
- PBX management
- voice mail management
- incoming call control
- router management
- WAN switch management

- network security management
- cable plant management
- technical management of telecommunications contracts
- cost control
- bill processing
- internal appropriation of telecommunications costs
- management of telecommunications contracts
- billing system management
- expense approval and budgeting
- documentation and administrative
- resource inventory control
- network diagrams
- administrative services
- internal telephone list
- planning and expansion
- capacity planning
- evaluating new technologies

The division of processes is arbitrary. Our only aim is to simplify the explanation. Also note that all processes are linked with each other. The objective here is not to describe all the processes but just to list them and give the reader a brief view of the ones associated with the telecom cost management, including which tasks they encompass. We will now discuss in detail cost control and documentation.

3.2.1   Cost Control

All activities linked with controlling the costs of the telecommunications infrastructure and the activities linked to internally appropriating this cost fall under cost control. It is important to emphasize the need for having a well-defined process to receive bills, match them with the actual usage and the resources actually in use, approve the payments, and properly register and distribute these costs internally. We are going to briefly discuss each one of them.

3.2.1.1 Bill Processing

This process encompasses all tasks related to receiving bills from the telcos (in electronic media or paper) and matching them with the resources actually contracted and the real usage. The organization should be able to process these bills in order to identify the expenditures by address, business unit, provider, and type.

The auditing of the bills is part of this macro-group of processes. It can be subcontracted to an external provider or made internally and executed on a regular monthly basis, or these audits may be executed on a quarterly or biannual basis. Although the auditing itself may not be executed monthly, the values verification and comparison between the values charged with the historical values needs to be done every month in a consistent way.

It is crucial to establish a process where every telecommunications invoice is verified by the telecommunications area before it is paid. This applies to all of the organization's telecommunications charges. Approval by the telecommunications area after verification and before payment prevents potential contract issues or missed steps in the process. The telecommunications area must be responsible for verifying both that the bill is part of a valid contract and that the value charged is actually due. Only then should the bill be forwarded to the accounts receivable department.

An effort should be made to consolidate the bills, so that each site does not receive its own bill. This consolidation should happen via electronic media. Telecommunications invoice, including telephone bills, should not be paid directly by the users. The telecommunications area must also be responsible for conducting the discussions about undue charges, penalties, and reimbursements.

Time limitations should also be included in contracts to limit disputes to a reasonable time frame (six months to a year), including verbiage related to service availability during dispute periods (this is a policy). We discuss this process in detail in chapter 9.

3.2.1.2 Internal Appropriation of Telecommunications Costs

This process encompasses the tasks linked to telecommunications cost appropriation to an organization's business units and cost centers.

In a telecommunications network, shared resources can make the proper cost appropriation far from simple. First, we have to clearly understand the concept of shared resources. The following example describes shared resources and some of the difficulties of appropriating their costs. As described previously, a connection may share ports, equipment, and last miles. A clear understanding of this concept is important because of the importance of the impact of cost appropriation and resource management. The following picture gives us a clear view of what happens physically.

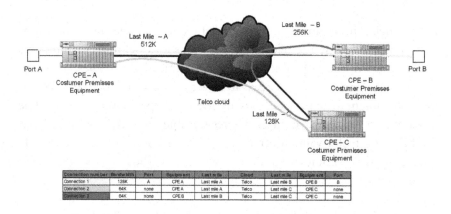

Connection number	Bandwidth	Port	Equipment	Last mile	Cloud	Last mile	Equipment	Port
Connection 1	128K	A	CPE A	Last mile A	Telco	Last mile B	CPE B	B
Connection 2	64K	none	CPE A	Last mile A	Telco	Last mile C	CPE C	none
Connection 3	64K	none	CPE B	Last mile B	Telco	Last mile C	CPE C	none

As we can see in the picture, the equipment and last mile A are shared for connections 1 and 2 (dark gray and light gray, respectively). This fact implies that the costs of these resources must be shared on a connection-by-connection basis: the cost to provide connectivity to each point of presence.

It is important to clearly define the sharing criteria before proceeding with a cost appropriation strategy. (Once again, it is an internal policy that has to be discussed internally by the organization.) For example, divide the monthly resource cost in the same proportion of the bandwidth used. In the case of last mile A, specifically, we have two connections: one of 64K and another of 128K. Using the criteria mentioned before, we would have a cost appropriation as follows:

Y = Last mile A monthly cost

Value appropriated to connection 1 for the use of the last mile A = 128 × Y/(128 + 64)

Value appropriated to connection 2 for the use of the last mile A = 64 × Y/(128 + 64)

64 and 128 are the bandwidths of connections 1 and 2, respectively, both using the last mile A.

The cost of the connection will be the sum of all entities associated with the connection (seven items).

This cost appropriation may also be mired in political issues within large diversified organizations when business units are themselves large organizations and pick services from central services that they perceive will directly benefit them. This may lead to different agendas on the part of business units, which detracts from aggregating costs to the advantage of the overall organization. This area needs careful consideration. Keeping the billing and cost appropriation as simple as possible should be a primary goal.

3.2.1.3 Management of Telecommunications Contracts

Telecommunications contracts need to be managed and their QoS and SLA levels verified. This type of activity demands time and effort. It is a mistake to assume that the existence of control tools and rigid QoS clauses in the contracts by themselves guarantee that failures will be identified and penalties applied. The processes must be not only documented, but also executed and audited to ensure effective execution. This process has a close link with the bill processing, given the fact that noncompliance with the SLA usually implies penalties or, in the very least, discounts. Therefore, it is necessary to identify breaches of the SLA when verifying the bills.

It is important here to emphasize the fact that management of contracts includes basic aspects, such as ensuring that the resource belongs to the contract, and complex issues like SLAs. As already mentioned, sometimes the perfect becomes the enemy of the good. When managing contracts, never try to implement complex controls before you are able to manage the basics well. You must be able to control the cycles of the contracts (following when each contract ends) and the resources they cover and their costs. Once you have these basic aspects well controlled, you can start controlling things such as historical monthly costs and reimbursements due to noncompliance with the SLAs.

3.2.1.4 Billing System Management

This process involves all tasks linked with managing, maintaining, and controlling the billing systems. It is crucial to maintain properly configured billing software and updated databases. Accurate information must reflect any changes in users, extensions, cost centers, the organization's hierarchy, tariffs, and least cost routes. These support tasks make the billing process effective.

To maintain accurate data, it is very important to establish processes that will keep the telecommunications area informed of new as well as departing users to the organization to facilitate timely action, for example, blocking extensions, mobile phones, and so forth. These are frequently localized HR processes, but a failure to link them to telecommunications departments may have cost as well as security implications.

2.2.1.5 Expense Approval and Budgeting

It is important to have a clear process where telecommunications expenses are only paid after the telecommunications area has approved them. Very often, organizations do not adopt this policy. It is very common to have the business units paying bills directly and only communicating the costs to the telecommunications area afterward, even that frequently isn't done.

This task of expense approval, billing, and cost appropriation has different levels of complexity, depending on the geographic spread of an organization's business units. Telco costs structures vary significantly between different regions of the world. For companies with a large international footprint, the picture becomes very complicated.

There has been a growing regulatory and organizational trend to emphasize financial transparency. Financial transparency has become an absolute must for senior management. This area is frequently woefully inadequate in large, decentralized organizations. Expense approval plays a pivotal role in financial transparency. When done correctly and following clear processes every time, expense approvals support very clear financial documentation and transparency. When not done appropriately, this leads to inaccurate budgeting and cost control, and, ultimately, a telecommunications environment that is not (or cannot be shown to be) running as an effective entity.

3.2.2   Documentation and Administrative

Documentation and administrative processes are associated with recording the details of the infrastructure, including inventories of business units, users, extensions, addresses, POPs, equipment, last miles, connections, contracts, configurations, and all associated administrative services.

2.2.2.1 Resource Inventory Control

In large organizations, telecommunications resources are often not well controlled. As time goes by, equipment and resources are relocated, deactivated, or returned to service providers without proper documentation. This leads to a very common situation where telecommunications management does not have an updated inventory of the resources under its responsibility.

To guarantee this inventory is updated, it is necessary that all people involved in resource management follow defined processes documenting any cancellations, changes, or moves. The telecommunications area must control these procedures directly. It is crucial that the information repositories be properly maintained and a process be in place to control this data. Controlling the data means not only keeping the depositories updated but also making sure the data is accessible, and having ways to transform this raw data into managerial information. Keeping the data updated is crucial but it is just as important to be able to access it easily and extract information from it.

3.2.2.2 Network Diagrams

This process includes all tasks linked with preparing and maintaining network diagrams. It is crucial to have proper network diagrams depicting the network as a whole and diagrams showing the configuration of every site and rack. The general diagram must show all the organization's POPs, all WAN devices, all last miles, and all connections. Such diagrams must allow the identification of every device under the responsibility of the telecommunications area and must be used as a day-by-day tool by the telecommunications team.

3.2.2.3 Administrative Services

A telecommunications area needs administrative support, including document forwarding, data entry, and so forth. There may be no need to have dedicated administrative personnel (the IT administrative pool may be enough), but depending on the functions executed by the telecommunications area, there may be a need for dedicated administrative personnel.

3.2.2.4 Telephone List Preparation and Distribution

This process includes all tasks linked with preparing, maintaining, and distributing the organization's internal telephone directory. Although the physical directory printing and folding may not be typically executed by the telecommunications area, its preparation has to be. Today, most companies do not even have a printed directory anymore; they rely exclusively on online directories. That is necessary in order to keep the directory in line with the moves and changes implemented in the PBXs and contracts.

Some departments update their own data in a common software directory. This strategy, although practical, tends to generate directories with a high level of mistakes, not in line with what was actually implemented by the telecommunications area in the PBX configuration.

In order to avoid such problems, operational uniformity must be guaranteed (administrative versus technical). It is crucial to have the telecommunications area in charge of controlling the organization's internal telephone directory. It is the only area that can ensure data integrity. The telecommunications area must have the responsibility for guaranteeing the uniformity of the information in the PBXs, voice mails, billing systems, and internal telephone directory.

This directory maintenance function will vary depending on the maturity of overall directory services in the organization and its PBX technology deployment. An organization with a mature identity management environment will typically have the phone directory integrated as a component of its overall directory and identity management implementation. In this case, the maintenance of the directory will be done by a different IT department. The telecommunications area will be responsible for its interface into the bigger directory environment, and the validity of information will be transferred.

3.3   Pitfalls to Avoid when Designing the Processes

Most telecom managers were telecom technicians or engineers before becoming managers; they usually understand the technical issues involved. That is, in general, a good thing. However, the fact that they know the subject profoundly doesn't guarantee that they will become good managers. Sometimes the ability to execute a task works against a manager. This happens for two reasons:

First, because the tasks are perceived as easy ones, managers may be reluctant to subcontract people to execute them, overloading their subordinates. This reluctance is based on a perception of how they would execute the tasks (usually in much less time than a median-level professional would take). Here it is important to keep in mind that the effort should be dimensioned not as if a highly qualified person was executing the tasks but planned to allow even median professionals to execute them. This is an important concept when designing the processes and defining the workload of the people involved. As the aphorism goes:

"The process must be designed by a genius to be operated even by idiots, not designed by idiots so that only a genius can make them work."

This is an important consideration because, as we all know, the structure should keep running as people enter and leave the organization, and you won't find a replacement if you need a genius.

Secondly, for the same reasons, managers may underestimate the difficulties that their subordinates have when executing tasks. Here we may illustrate the point using another aphorism:

"Everything is easy as long as you know how to do it."

The problem is, sometimes even the most obvious thing is not known by the person executing the task. It is difficult to understand that some people simply don't know things that others consider as basic. How many times do we hear comments like "How could this guy not know that?" When designing the process, you shouldn't presume the people executing them are experts or have even good sense. Even the obvious alternatives and explanations should be mentioned, with details.

Here is another important aspect that is very often overlooked: put too many simple things to be done together, and they will become very complicated. This concept should be kept in mind when designing the process; what makes a process difficult isn't the fact that the individual tasks are difficult, but the fact that when executed together and at the same time, they become complex.

Technical staff newly promoted to management very often don't understand that their objective is to get results through people. Sometimes a technical manager thinks that it is easier to just do a task than delegate it to someone (who may need training and supervision).

Organizing the processes is in essence a strategy to train people in how things should be done, when and by whom, and of course taking how long. The management must understand the importance of this. It is pretty obvious but often forgotten: you can't do everything even if you know how to do it.

3.4 Definition of Tasks and Responsibilities

A clear definition of tasks and responsibilities is crucial to having well-organized processes. It is important to define clearly who does what, when, and how. For example:

When organizing the process of treating the bills, it is crucial to define your terms. Who receives the bills? When do they usually arrive? How are they received (e.g., downloaded from the telco website)?

Once the bills are received, how are they treated? Who registers the value charged and how (by bill, by trunk, by contract, etc.)? How are the bills verified? How do the people in charge of verifying the bills get them?

Once the bills are verified, how will you register the values to be charged? How will the discrepancies be treated? How are the interactions with providers supposed to happen? How will the paid values be registered? Which reports will be generated, with which frequency, by whom and to whom?

The clear answer of all these questions is what allows the defined process to really work. Of course, you can't think about the tasks and responsibilities detached from the processes and policies. They are two faces of the same coin.

3.5 Norms and Procedures

The proper management of a telecommunications area demands a serious organizational effort. For a telecommunications area to operate properly, it needs adequate operational procedures and norms. Those depend on the context of each organization. The telecommunications managers must detail them for their specific environments. When defining the procedures (derived from the processes), we should indicate the responsibilities and frequency, and, above all, identify who can do what and when; for example, Who can request a specific report and how should this request be done?

The tasks involved must be identified, as should the necessary skills required of the professionals who execute these tasks and the workload involved. Such task mapping allows the precise definition of the necessary workforce whether the whole operation is in-house or outsourced.

The task mapping must include the identification of the tasks the organization intends to execute in-house and the ones that will be outsourced. The job descriptions of the staff members must indicate their tasks and responsibilities and the workload involved. This enables the identification of under—and overallocation of personnel as well as the identification of technical deficiencies (if any) solved through training and redeployment of people.

The goal is to provide the organization with all the necessary instruments required to effectively manage the telecommunications infrastructure. Such instruments include an appropriately sized, trained, and organized team (where all members know their responsibilities and are well prepared to execute them), clearly defined processes and policies, and complete infrastructure documentation.

3.6 How to Allocate the Control Effort

As in every human field, in telecom management, sometimes perfection is the enemy of the good. This is why it is very important to focus on the basic information to be controlled. You have to define what you consider basic, what you consider important, and what you would like to have if you could. In other words, you have to define the list of information that you must have, the information you could have, and the information you would like to have. This division is important because the resources to control the infrastructure are usually limited and you have to make hard choices. But what is basic? The answer varies; however, here are some good guidelines:

- the location's inventory
- the organization's business unit inventory the
- hardware inventory (including the telco devices installed within your premises)
- the service provider's inventory
- the data circuits inventory
- the voice access inventory
- the contracts and tariffs inventory (current and past)
- Total expense with telecom and the total cost broken by provider, contract, location in a per month basis (historical costs).

This is the basic information you must have. You must have it readily, it must be updated, and it must be reliable. A frequent problem used to be balancing the effort, with some processes expending too much time to control things that are not so important or which could be controlled with fewer details, generating equally good results.

For example, you may decide to manage your billing system using passwords instead of billing the calls by extensions. Of course, the control using passwords has several advantages; however, it obligates you to keep the user/password database updated and requires additional work to maintain. Depending on the context, the additional work may not justify the gains in terms of control.

We should be vigilant with regards to the processes whose cost/effort surpasses the associated economic benefits. We should always keep in mind that in most processes, the biggest costs are not the initial implementation but the maintenance. Most telecom teams are small and overworked, and their responsibilities encompass not only tasks associated with control of resources for the organization as a whole but also operational activities associated with keeping the lights on. Then it is important to avoid drawing away the focus of these people to secondary tasks whose benefits are small and specific.

For example, an organization has a billing system that collects logs of 61 of its 294 sites; the maintenance of this control process demands a disproportional amount of time, which benefits only a small part of the organization. The telecom manager has to understand this kind of problem and maintain a balance between the available resources and the focus of the control.

Again, the focus has to be in the general control of the costs, coming from the macro to the micro, as the time and availability of resources allows. It doesn't matter how much a specific site spent if you don't know the telecom cost of the organization as a whole. Although it seems a bit obvious, this is a very common mistake. The costs must be controlled from the macro to the micro, never reversing this sequence.

Another important aspect telecom managers must attend to is the fact that third-party providers tend to focus on their own processes and draw the organization toward their own needs. For example, the billing provider tends to emphasize the importance of the processes associated with the billing system; the bill auditors will do the same to the processes associated with auditing bills; and so on. It is up to the telecom manager to put the priorities right.

For example, processes like control of historical costs, invoice management, and resource inventory usually rank way above billing systems and bill auditing.

Very often, professionals linked to billing control and billing systems refer to their jobs as telecom expense management; in fact, their jobs are part of the TEM process, not the process itself.

In summary, it is crucial to deploy the scarce resources wisely, always comparing the benefit of a control process against its cost in terms of time and effort.

Chapter 4: Cost Benefits of Expense Management Methods

When defining strategies for controlling telecom resources (including costs), it is always important keep in mind that we have to balance the effort needed to control these resources with the benefit of the control. Not every control is worth the effort, and even when the effort to control pays off, it may only be the case if we properly define the granularity of the control.

Here, more than anywhere else, the motto "The perfect is the enemy of the good" is applicable. The telecom manager has to evaluate each control process, identifying carefully what should be controlled and how, including the granularity of the control. Therefore, when doing this kind of evaluation, a crucial parameter is the percentage of potential gain or loss if the control isn't in place.

For example, when analyzing the process of billing auditing, some decisions are important:

1) How will the auditing be executed: call by call or by grouping types of calls?
2) How frequently will the process be audited: every month, more sporadically, or only when an untypical value is spotted?
3) When spotting discrepancies, how will you proceed: pay and ask for reimbursement, hold the whole payment, or pay part of it?

All these points demand an understanding of the benefits as well as the effort associated with each strategy.

A decision-making process like that must be conducted in all efforts associated with controlling the telecom infrastructure, always evaluating the tradeoffs between the effort to control and the benefits (or potential loss) associated with that control.

Here, it is worth mentioning that some evaluations are more difficult, because the benefits are not so clearly measurable. For example, some benefits are operational, not economic, and there is no direct translation of these benefits into economic terms. For example, how do you represent in economic terms the productivity gains associated with knowing exactly where each cable is in the patch panels? We know that there is a productivity gain for the technicians maintaining these patches; however, it is hard to translate these gains in economic terms and hard to weight the control effort against the benefits.

In this context, we see the importance of using tools that minimize the control effort. Simple things like having only one data depository, accessible to all, can have a big impact in terms of control effort. Guaranteeing that data is recorded only once and is available to all once recorded generates savings in the control effort.

Another point to be observed is that the effort spent in recording the data should not jeopardize the effort dedicated to transform this data in real understanding of the phenomenon. This is an important point. Any process must include both activities linked with producing the information and activities associated with analyzing it. A typical example of what we mean by that is when an organization produces several cost reports without any critical analysis of why the costs went up or down.

In the same way, time must be allocated to discuss the analysis internally and make understand it vertically as well as horizontally.

The point here is to identify the equilibrium between effort and benefits of the controls. Therefore, we can put the issue as follows: The full benefit of the control is only achieved if the information generated is properly analyzed and spread throughout the organization. Therefore, when evaluating the cost benefit of a control effort, it is crucial to include not only the effort to produce and record the information but also its analysis and diffusion.

A control process that contemplates only the production and recording of the data but doesn't include analysis will not produce complete understanding of the phenomenon. By the same token, if we produce the right analysis of the phenomenon but are unable to spread this understanding throughout the organization (vertically and horizontally), this understanding will not translate into effective action; it will be sterile. Therefore, when measuring any control effort, we must include all three phases (control, analysis, and diffusion); the joint effort of these three phases should be compared with the benefit. A cost benefit analysis that doesn't include all three phases minimizes the real control effort.

Such considerations are applicable from basic control processes to sophisticated traffic analysis. For example, even if we have an adequate inventory of contracted resources, it is worth nothing if they are not associated with the bills. Only this association allows verification of whether the resources are being charged properly. In the same logic, even if we can compare what is supposed to be charged with what is actually being charged, if there isn't a mechanism in place to inform the payment area about the discrepancies, the knowledge of the discrepancies will not generate any action (e.g., keeping the organization from paying for undue charges). As we can see, even in a very simple process like that, we need to have three phases: control, analysis, and diffusion.

Organization size should also be considered when defining control granularity and effort. Here, it is interesting to note that the size of the telecom team doesn't always grow proportionally with the size of the organization; this fact imposes a logic where the bigger the organization, the less granular the control implemented. This is a factual finding. Therefore, if you are a telecom manager for an organization with two hundred sites and a team of four, where even the cables in the patch panels were controlled, and you move to a new organization with fifteen hundred sites and a team of eight, you should reevaluate the granularity of control you want to see implemented, guaranteeing an adequate tradeoff between effort (and availability of people) and benefits. You must guarantee at least a general view of the cost of the infrastructure and the traffic patterns. In sequence, you should guarantee an adequate maintenance in the resource and contract inventory and control all interactions with providers.

In this context, it is very important to keep in mind that sometimes small changes in the process may have a big impact in the control effort.

For example, if we audit the bills every month, we may see a scenario where if we have some problem with the values charged, the payment isn't executed and a new invoice is requested from the provider. If we change the control process so the bills are accumulated for six months before being audited we reduce the effort without significantly changing the values spent by the organization. This happens because the effort to audit and ask for reimbursement for six bills is similar to the effort to do it for one bill. In addition, auditing the bills every six months dissociates the approval/payment process from the auditing process. That means the auditing doesn't have to be done before a payment is approved (this eliminates critical dependence). This change also makes it easier to outsource the auditing process, given the fact that we don't have a mandatory time span for the conclusion of the work. The downside is the fact that we have to pay the invoices in full, even if we suspect that there are mistakes and reimbursement may take longer.

Chapter 5: Asset and Service Inventory Management

At the risk of stating the obvious, a crucial aspect when managing a wide area network is controlling the inventory of resources. Effectively controlling telecom costs requires precise information about the following entities:

- the organizations and business units served by the network
- the physical locations (addresses) touched by the network
- the points of presence of the business units in the physical locations
- the service providers supporting the organization
- the contracts with each service provider, including the price lists of each contract and the resources supported by each contract (including the tariffs applied to each specific type of traffic)
- the equipment used by the organization, including those that belong to the organization and those rented or leased, including their location and cost
- the last miles used by the organization, including information such as capacity, location, and cost (the term "last mile," although more often used for data access, also applies to voice trunks)
- the dedicated connections used by the organization, indicating the capacity and location of each end (A and B)
- the control of all interactions with the service providers: requests for new resources, cancellations, changes, and maintenance (every interaction must be documented and traceable)

- the monthly cost of the infrastructure per organization, business unit, service, service provider, contract, address, point of presence, and individual resource (equipment, last mile, and connection)

Managers should have this information, but that doesn't necessarily mean that they will have a system to control it. Such inventories and controls can be entered in a spreadsheet (the benefits of using telecom expense software are discussed further in this chapter). The key point is that if telecommunications network managers cannot find this information quickly and accurately, they cannot manage the structure effectively.

It may sound strange, but the people who are supposed to manage the network often don't have a complete view of the resources under their responsibility; this is common in large multinational companies where the structure is contracted by different groups within the organization (for example, part of the network may be contracted by the headquarters and other parts controlled by the local teams). To add to this problem, it is very common to see different business units contracting telecommunication resources, sometimes not even bothering to inform the people who are supposed to manage the whole structure. This often leads to situations where managers, due to either management failures or cultural/systemic situations, don't effectively execute their mandate.

The existence of a network operational center or monitoring tools does not, by itself, guarantee an adequate inventory of resources; above all, it doesn't guarantee an adequate link between resources contracted and values expended (expenses or depreciation). It is absolutely crucial to build and maintain a solid link between the physical and financial aspects of the network.

5.1 The Meaning of Each Controlled Entity

We are now going to standardize the definitions of the entities that should be controlled. Understanding these concepts will make subsequent topics easier to comprehend. Here, it is interesting to note that although most of these concepts may sound extremely simple and obvious, when combined they may not be. In addition, we must understand them well enough to be able to recognize them even where the nomenclature is different. We can divide the entities to be controlled into three types:

- functional
- technical
- structural

Note that this classification is arbitrary and is only used to make the process easier to understand. Our objective here is to give you a structured way to control your data, even if you only use a spreadsheet; we strongly suggest you organize your data as described here.

5.1.1   Functional

Functional entities are associated with the network but are not components of the network itself. Here are eight functional entities:

- **Organization:** This is the entity that encompasses the business units. Here is a practical way to visualize what that means: think about a state government as the organization and its several departments and state companies as its business units. The same concept can be applied to a large multinational corporation; its several country operations and business segments can be considered independent business units.
- **Business unit:** This is a subset of an organization whose criteria for segmentation can be multifold (geographical, type of activity etc.).
- **Addresses:** These are all the physical locations where the organization has telecom resources (in other words, all addresses touched by the organization's network). Note that the address may not belong to the organization; it may be a client's site. If there is a telecom resource in this location that belongs to the organization, this is a point of presence, and consequently, this address has to be identified.
- **Points of presence:** These are the physical locations of the business units; note that the same address may have several points of presence for more than one business unit.

- **Service providers:** These are the companies providing telecommunications services for the organization.
- **Contracts:** These are the agreements between the organization (or its business units) and the service providers.
- **Price list:** Each contract has a price list associated with it.
- **Requests:** These are interactions between the organization (or its business units) and the service providers. The organization can request installation of new telecom resources, changes, cancellation, or maintenance of the existing ones.

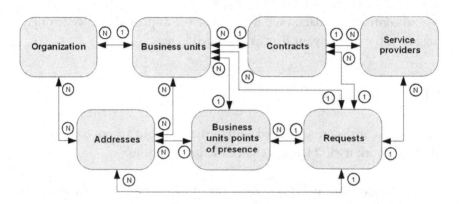

The following list describes some relationships among these entities:

- One organization may have several business units.
- One business unit belong to only one organization but could be present in several addresses (points of presence).
- An address can host several business units (one point of presence for each business unit).
- A point of presence belong to only one business unit and is located in only one address.
- A business unit can have several contracts with several service providers.
- A business unit can make several requests to several service providers.
- A service provider can have several contracts with several business units.
- A service provider can provide several last miles, equipment, and connections for several of the organization's business units.

- A contract can belong to only one service provider.
- A contract can belong to only one organization.
- A contract can support several business units within the same organization.
- A request is linked to only one contract (although there can be several requests associated with one contract).
- A request can be associated with only one contract (and consequently with only one service provider).

5.1.2 Technical

- last miles (data accesses, Internet accesses, voice trunks)
- equipment and ports (customer premises equipment, PBXs)
- service provider network (represented by the cloud)
- connections (PVC, CVP, channels, traffic, etc.)

The following picture represents the relationship among these four elements:

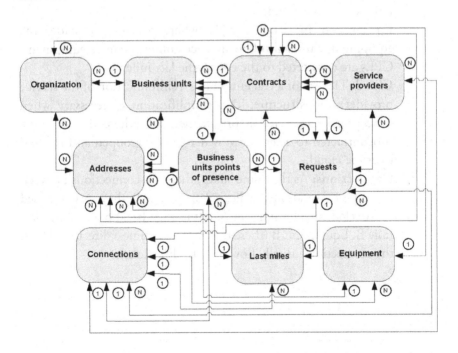

As we can see in the picture, a connection between any given location A and any other location B is composed of six elements:

- end A equipment, with or without an associated port (CPE A)
- end A last mile, which is the local loop between the address A and the service provider backbone
- service provider backbone (represented as a cloud)
- end B last mile, which is the local loop between the address B and the service provider backbone
- end B equipment, with or without an associated port (CPE B)
- the connection between A and B (usually called PVC or CVP)

Such elements belong to four types of technical entities: two last miles, two equipment, the cloud, and the connection itself. Here are the four technical entities:

- **Last miles:** These are the local loops connecting the address to the service provider backbone (the provider backbone is usually represented as a cloud).
- **Equipment:** In data networks where the telco provides the equipment, this is also known as customer premises equipment. CPEs are connected to the end of the last mile.
- **Cloud:** This is not a real entity; it is a simplification of the service provider's infrastructure. This simplification is necessary when conceptually analyzing a private network where the resources subcontracted to a service provider can be grouped and referred to as the cloud.
- **Connections:** As the name states, these are connections between two physical locations (dedicated or switched); they are composed of five elements: Equipment A, last mile A, backbone (cloud), last mile B, Equipment B (these terms don't assume a directional bias; the connections usually flow in both directions).

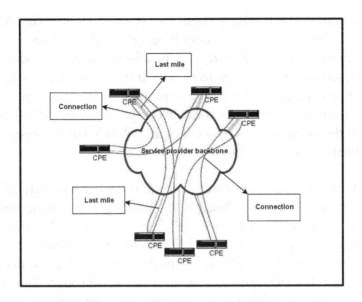

The relationships among the entities shown are as follows:

- A service provider can have several contracts, each one with several last miles, equipment, ports, and connections.
- A contract can include more than one last mile, equipment, port, and connection.
- A connection can be composed of last miles and equipment belonging to different contracts and service providers.
- Each last mile, equipment, port, and connection belongs to only one contract.
- A last mile and equipment can be used by several connections.
- A last mile connects the service provider's backbone to only one piece of equipment, located in only one address.
- A connection has only one piece of equipment and only one last mile associated with each one of its ends.
- A connection is associated with only one point of presence (and address) in each one of its ends.

It is interesting to note that equipment and a last mile can be shared by different connections. It isn't uncommon to have equipment and last miles with one, two, three, or more connections. Each connection, though, is associated with only one equipment and only one last mile in each one of its ends.

Every connection encompasses eight cost factors (Equip A/Port A, Last mile A, backbone, last mile B, equip B/port B and the connection itself). Depending on the pricing strategy adopted by the service provider, some of these cost factors may seem to be zero. It is also possible that in cases where the equipment is owned by the organization, this equipment still has a cost, which is represented by the total cost of ownership (TCO), which includes the weighted average cost of capital (WACC) of the organization plus maintenance and management costs.

The idea is to identify the cost of providing connectivity between two defined points, A and B. Conceptually speaking, every connection has these cost factors associated with the technical entities. The equipment and last mile costs are easily linked with the cost of providing connectivity between point A and B (although sometimes shared by more than one connection). The cloud cost, though, is shared by all connections, which makes it a bit more difficult to appropriate costs properly to a specific connection.

For clarification: The term "connections" refers to entities representing throughput between two points. Throughput has two basic attributes:

- minimum guaranteed flow (usually in packet networks referred as CIR)
- maximum possible flow (usually in packet networks referred as EIR)

This concept isn't linked to the technology used to provide the connection. In summary, the organization buys throughput among its points of presence from the service providers. The picture shows the eight entities that constitute a connection:

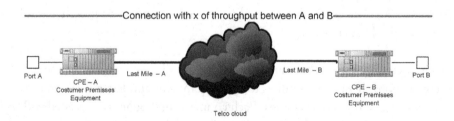

One element not shown is the traffic. This is just a basic conceptual model, and depending on the pricing model adopted by the service provider, some items may not be present or may not have costs associated with them. We use this framework just to understand and normalize the different pricing strategies.

For example, a given telco may not charge for the ports and equipment, charging only for the last miles and for the PVCs (the price varying with the bandwidth). Other telcos may charge for the last miles (the price varying with the bandwidth) and equipment (fixed price), not charging anything for the PVCs. The point is that different services (and service providers) appropriate costs differently; this framework allows us to normalize this cost appropriation.

5.1.3   Structural

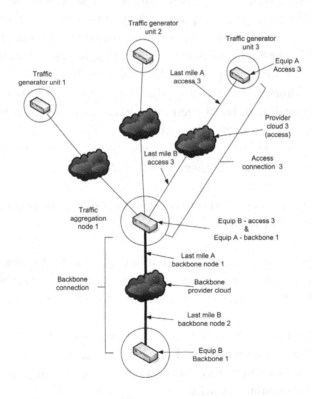

The picture above shows how the technical entities are articulated in a non-star WAN. This diagram is a bit more complicated than when we have a pure star topology. Here, we have a situation where there are aggregation nodes and different service providers for the connections. In addition to that, we have connections linking the sites to the nodes (access) and connections linking the nodes between themselves (backbone). This picture helps us understand structural entities:

- **Traffic generator unit:** In this context (corporate network), traffic generator units usually are workstations and people (although some variation of this basic concept may exist).
- **Traffic unit:** These are groups of traffic generator units; in a corporate network, these units are usually the organization's points of presence (in a call center, these may be users grouped by area code).
- **Traffic aggregation node:** This is where we aggregate the traffic before distributing it (for transportation reasons or for final destination). Following an analogy with a logistic network, the nodes would be a warehouse where the goods are stored and from where they are distributed. A given location can be a traffic unit and a traffic aggregation node at the same time.
- **Backbone:** This refers to the connections between the traffic aggregation nodes (the term may have other meanings, but this is what we mean throughout this book).
- **Access:** This refers to the connections between the traffic units (sites) and the traffic aggregation nodes (nodes).

Backbones are not associated with bandwidth (although there is a tendency for backbone connections to be bigger).

Another aspect that is not always obvious is the fact that equipment and last miles can be part of access and backbone connections (for example, Equip node 1 in the picture supports connections to the node and to the sites).

It is important to understand that such concepts are applicable to both private and public networks (dedicated or switched). In this book, our focus is on private networks; some concepts may be slightly different when applied to public networks.

The subsequent pictures illustrate how the traffic generator units are associated with the traffic units. As we can see, in a pure data network, the traffic generator units are workstations (and, of course, servers); in a pure voice network, the traffic generator units are people (extensions); in an integrated network, they are both.

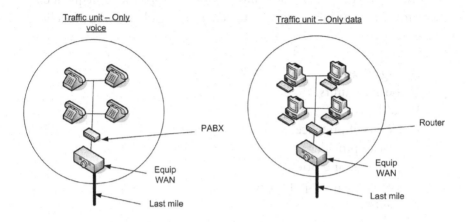

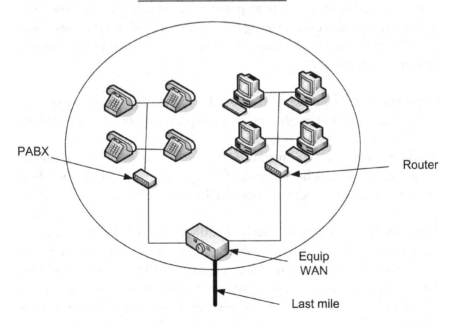

Traffic unit - voice and Data

5.2   How to Map the Information

When building an inventory of telecom resources, besides understanding the meaning of the controlled entities, it is very important to map the information following a rational sequence. Following this sequence is important, because each layer of information supports the subsequent one. Therefore, it is crucial to organize the data to be gathered in a predefined sequence:

1) organization
2) business units
3) addresses
4) points of presence
5) service providers
6) contracts and price lists
7) last miles and equipment
8) connections

We already defined the meaning of each entity; now we are going to show that when gathering the data to populate the databases, it makes sense to proceed in sequence, building layer over layer of information. Now we will discuss each term with an emphasis on how to build a proper inventory of resources:

Organization: The organization is the entity that encompasses the business units; this a straightforward concept and doesn't need discussion.

Business unit: As explained previously, a business unit is a subset of an organization whose criteria for segmentation can be multifold (geographical, type of activity conducted, etc.).

Note that once you have clearly identified the organization and the business units, it is time to identify the addresses and points of presence. These two concepts may sound equal, but in fact they are not. An address can be the point of presence of several business units. For example, in a large organization with two business units, you may have an address where both business units are present, and therefore you have two points of presence within a single address.

It is important to remember that the objective is to map the location of the telecom resources. Therefore, we often have a situation where the resource is installed within a client or provider address; in these situations, the client's address has to be identified, and it will be considered as a point of presence of the organization within the client or provider. In other words, addresses are all locations where the organization has telecom resources (by resources, we mean assets whose financial burden is supported by the organization (even when installed outside its premises).

As already mentioned, it is highly advisable to follow a rational sequence when mapping the infrastructure; each layer of information gathered helps to build the subsequent one. For example, if we map the addresses properly, we cannot fail to ask which telecom resource is installed in each address (identifying situations such as when the site doesn't have a PBX). Proceeding in layers like that helps to ask the right questions and avoid mistakes.

In terms of organizing the telecom contracts, the following sequence is recommended:

1. service providers
2. contracts: price lists

Service providers: As explained previously, these companies provide telecom services for the organization.

Contracts: Contracts are the agreements between the organization's business units and the service providers. In general, we usually consider each bill as a consequence of each contract. That means if you receive ten bills per month, it is as if you had ten contracts. Of course, you may have only one real contract, to which all ten bills are associated; however, in terms of building a database of your telecom resources, it would be advisable to match the number of bills with the number of contracts (even if they are not exactly correlated). This is because you must be able to identify the number of bills you have to pay each month, and if any bill doesn't appear, you must be able to take notice.

Each contract must have an associated price list; each price list may have several items to classify:

- last miles
- equipment
- ports
- traffic
- cloud
- connection

To properly document the price list, you must understand the rules applied when defining the transport prices. This understanding requires identifying the parameters and factors that the service providers use to determine their prices and correlate them to the six predefined possibilities described above.

Being able to put all contracts in the same framework is important; it is also trick. It is important because it makes you identify your price structure throughout a multitude of services, service providers, and countries; it is trick because the charging strategies can be very opaque. In general, telecom charging strategies oscillate between the following two extremes:

- 100 percent predefined values per services
- 100 percent variable values (depending on actual usage; for example, spoken minutes)

Although most services are charged somewhere in between these two extremes, data access usually has its price 100 percent predefined; there are usually five cost components, plus taxes and discounts:

- Last miles usually have their prices defined based on the nominal bandwidth; associated with last mile in our framework.
- The maximum amount of ports and type of services supported usually defines the price of the equipment; associated with equipment in our framework.
- Ports usually have their prices defined based on the speed (bandwidth). We may bundle the port costs into the equipment cost or associate it separately.

- Management is usually a defined value for the network as a whole. This value is sometimes divided by the number of circuits and charged on a per-circuit basis. Usually we bundle this cost into the equipment or into the last mile cost.
- The connections prices (whether PVCs, CVPs, or point to point) usually are defined in terms of CIR, EIR, or nominal bandwidth; associated with connection in our framework.
- Taxes vary based on the country, state, and even city where the connection is installed.
- Special discounts can be based on specific negotiations.

These cost components sum to the total cost for data access service; providing data connectivity between two points usually implies the following cost structure:

Port A + Equipment A + Last mile A + (PVC/connection/cloud) + last mile B + equipment B + ports B) = Total cost plus taxes and discounts (consider management bundled into the last mile or equipment cost).

Voice services usually are priced in a mix between 100 percent predefined and 100 percent variable; there are usually only three cost components defining the cost:

- Last mile is usually a monthly subscription fee; associated with last mile in our framework (100 percent predefined).
- Usage is usually the value charged that is associated with a volume of spoken minutes. The cost of the spoken minutes may vary depending on the distance, hour of the day, area code, national/ state borders, or day of the week. Sometimes, a minimum amount of traffic is charged, whether it is actually used or not (some charging strategies are on a per-call basis, regardless of duration). This cost item is associated with traffic in our framework. We may identify the several types of traffic; for example, long distance, local, or mobile. In our price list, we identify the unit and the cost per unit (for example, one minute of long distance call: USD 0.01.
- Taxes normally vary, depending on country, state, or city where the connections are installed.
- Special discounts can be based on specific negotiations.

The formula for the total cost of voice services is therefore the sum of only three components:

Last mile A + volume used × cost per unit = Total cost plus taxes and discounts.

Although we may find variations of these basic components, in a vast majority of the cases, the prices will be defined based on this cost structure. Knowing how the tariff is structured is important to understand how the total cost is built.

Sometimes, the cost appropriation does not use all cost components. Often, this cost appropriation is divided between the two ends of the connection. For example, in a given phone call, call originators usually pay for the subscription of their own trunk (last mile) and for the traffic. Users receiving the call also pay for the subscription of their own trunk. In a situation like that, the match with the cost components described previously is as follows:

- **Equipment and port A:** Cost appropriation is zero, but this may not be true in some cases, such as when the telco provides the PBX (called CENTREX services).
- **Last mile A:** This is the trunk subscription, which the call originator (caller) pays.
- **Cloud—the telephone network:** The cost appropriation is zero.
- **Last mile B:** This is the trunk subscription; the receiver pays for this.
- **Equipment and port B:** The cost appropriation is usually zero, but this may not be true in some cases, such as when the telco provides the PBX.
- **Connection and traffic:** The caller usually pays for the call based on the duration.

Chapter 6: Service Ordering and Change Control

This chapter will discuss the processes associated with requesting new resources and canceling or changing existing ones.

Here we have to understand that the process of requesting a new resource rarely occurs through a retail negotiation; it is usually done within an existing contract, where the price is already defined.

This chapter is more concerned with the mechanics of requesting new resources and canceling or changing the existing ones.

The definition of these mechanics is very important, because it has a direct impact on the processes of controlling the resource inventory, checking the bills, and controlling the contracts. To be able to affirm that a resource should be paid, you have to know who asked for it, when it was installed, and how much it was supposed to cost (which contract and which item in the price list it is connected to).

Besides, a new resource implies additional costs (or a canceled one in a reduction of charging values), so it is absolutely critical that each request made to the service provider is formal and controlled.

Requesting new resources, and canceling or changing existent ones, needs to follow a predefined process. This process must include the following phases:

1. Formal initial request: This phase implies that the organization asked the service provider for one of the three options (new, cancel, or change); the service provider must answer this request within a defined time span. Note that "answer" here doesn't mean actually do what was asked; the provider may just inform you whether what was asked is feasible, within which time, and how much it would cost. We generally call this phase "waiting for the service provider's answer." Discussing how much the resource would cost constitutes a retail negotiation, and unless you already have the item in the price list of an existing contract, you have to do that for each resource requested.

Here it is worth mentioning that in some organizations, the quotation process is separated from the actual ordering. In this scenario, the order of a new service only happens after quotations from many providers have been received. In this context, the formal initial request has the objective of defining the delivery time of the resource. This strategy implies that nothing is asked of the provider until the price is already known and a contract already exists.

2. Formal answer from the service provider: This phase encompasses the answer of the service provider, saying whether it can execute what was asked, how long it would take, and how much it would cost. Then the process turns into waiting for organization approval.

3. Approval (or not) by the organization: Once the service provider determines if it can execute the request, within which time span, and how much it would cost, the organization can evaluate whether it still wants what was requested (maybe the cost is too high or the time span too long). If the organization doesn't approve the request, the process is finished. If the organization approves the request, the next phase becomes "waiting for service provider implementation."

Note that you must have some mechanism to control the time between each phase; this time is usually set in the contracts as part of the service level agreement (SLA). For example, if you have five days to the service provider answer and five days to organizational approval, you need to be able to control the request status. Whether it was answered within the defined time or not, the same control is necessary for the organization approval.

Of course, you must be able to control the schedule given by the service provider to execute what was asked. For example, if the service provider answered that it can install a new last mile within twenty days, you need a mechanism to follow the request, indicating whether it was achieved or not.

4. Service provider informs the organization that the request was concluded: After the service provider finishes what was requested, it must inform the organization. Once this is done, the phase becomes "waiting for organization to test."

5. Organization tests whether the request was fulfilled: The organization must have a deadline to test and accept (or not) that the request was executed. If the request is accepted, the request is complete, and the financial implications become active. If the organization doesn't accept the request, it returns to the previous phase: "waiting for service provider implementation." Here it is worth including two caveats: 1) A mechanism usually exists where if the organization defaults, the time to test the request is considered finished. It works as if the organization had waived the right to test. 2) The request can return to the previous phase as many times as necessary, until the organization considers that what was requested was delivered.

When we mention that the financial implications of the request were activated, that means the cost of the new or changed resource should be automatically included in the contract (or the cost of the canceled resource should be removed from the contract and consequently from the bill).

The four phases described apply to requests for new resources and to changes and cancellations of existing ones.

6. Maintenance requests for a telecom resource to a service provider requires a specific workflow. The main difference between the maintenance workflow and the general requests (new, cancel, or change) described in the previous paragraphs is that there is no need for the service provider to inform the organization whether it can execute the request (or the time span or price). That information is all predefined. In a request for maintenance, everything works according to the phase, "waiting for service provider implementation.

In summary, requesting services from providers demands a set of rules and procedures that basically following this sequence:

1. The organization makes a request.
2. The service provider answers how much it is going to cost and how long it is going to take to execute (or it may say it can't execute the request).
3. The organization approves the price and time frame (or not).
4. If the organization approves, the service provider starts executing what was requested, and the time countdown starts.
5. When the service provider finishes, it informs the organization that the request was concluded.
6. The organization checks to determine if what was requested was properly executed and approves it (or not).

If the request is for maintenance, the sequence is slightly different, in that some phases are not necessary. In a maintenance request, the organization doesn't have to approve the execution of the request, and the service provider doesn't have to inform the organization of the time needed to execute it (that is defined in the contract as part of the SLA).

The exception for this rule is a maintenance request for equipment without a maintenance contract; in this situation, all steps have to be followed.

A request is the stage where several actors of the process of requesting and executing the four types of tasks interact. Therefore, each one of these actors must fulfill (or "edit") the request as it goes through the process. Service providers have to be able to inform the date when they believe the request will be concluded, and then the organization approves it (or not,). If it was approved, the service provider will inform the organization when the service is ready, and so the process goes. Each actor has to fill its respective field as the process continues.

A telecom management tool makes it very easy to link these actors, phase by phase. Of course, you may be able to manage this process without deploying such a tool, but it can be very hard to follow all these steps and phases using spreadsheets and e-mails.

The request process (not including maintenance) has four phases; there are two possible statuses in each phase: on-time and delayed. The maintenance request process has only two phases and two statuses: on-time and delayed.

A request for a new resource or a request to change or cancel an existing one can have the following phases/status:

• Waiting time definition by the service provider	On-time	Phase 1
• Waiting time definition by the service provider	Delayed	Phase 1
• Waiting confirmation by the organization	On-time	Phase 2
• Waiting confirmation by the organization	Delayed	Phase 2
• Waiting installation by the service provider	On-time	Phase 3
• Waiting installation by the service provider	Delayed	Phase 3
• Waiting test/approval by the organization	On-time	Phase 4
• Waiting test/approval by the organization	Delayed	Phase 4

This sequence presupposes that there is a service level agreement in place defining the four time limits, which separate on-time from delayed:

- **Time limit 01**: Maximum time that the service provider has to inform the organization how long it is going to take to execute the request. **"Waiting for service provider answer."**
- **Time limit 02**: Maximum time that the organization has to confirm (or not) the request based on the **time to execute** proposed by the service provider. **"Waiting organization approval."**
- **Time limit 03**: Maximum time to execute the request, informed by the service provider when it answered the request, and approved by the organization when it confirmed the request. **"Waiting execution by the service provider."**
- **Time limit 04**: Maximum time that the organization has to test and approve (or not) the service executed once the service provider

informs the organization that the service is ready. **"Waiting for test/approval by the organization."**

Once the request is generated by the organization, the service provider has a given number of days to answer it, stating how many days will be necessary to execute the request. The number of days to answer is defined in the contract.

Once the service provider informs the organization how much time is needed to execute, the organization has a given number of days to confirm (or not) the request. If you don't accept the time proposed by the service provider to execute the service, the request dies (you still must inform the provider that you did not accept the proposal).

Approving a request has economic and operational impacts, and therefore only authorized people should be able to do that.

After the organization accepts the proposed time to execute and approves the request, the time for installation starts counting. The service provider will be evaluated as to whether it succeeded in delivering the service within the time of its own defining. Therefore, in this context, "delayed" means that the service provider failed to deliver the request within the time it proposed for the organization.

Once the execution of the request is concluded, the service provider must inform the organization. Once informed, the organization has a given number of days to test and accept (or not) the service. The number of days to test and accept (or not) should be defined in the contract.

If the organization tests the service executed and does not approve it, the request returns to the status "waiting for installation by the service provider."

If the organization does not test the service executed and doesn't approve it within the time defined, the request stays as "waiting for test/approval by the organization" but "delayed." It may be agreed with the service provider that resources not tested within a given number of days are considered accepted by default.

In its turn, maintenance requests don't have to be answered or approved; they go directly to the "waiting for execution by the service provider":

1.	Waiting for installation by the service provider	On-time	Phase 3
2.	Waiting for installation by the service provider	Delayed	Phase 3
3.	Waiting for test/approval by the organization	On-time	Phase 4
4.	Waiting for test/approval by the organization	Delayed	Phase 4

As already mentioned, the maintenance request doesn't go through phases 1 and 2, since it isn't necessary for the service provider to determine the time to execute (or for this time to be approved by the organization). For maintenance requests, this "time to execution" was previously defined in the contract. The number of hours to solve the problem is defined in the contract.

Once a maintenance request is made, immediate action is required from the service provider. The service provider only has to inform the organization when it solved the problem, in order to allow the organization to test it and then accept that the problem was solved (or not).

6.1 Unifying the Communication Channels with the Providers

In large organizations, several people usually interact with their providers, asking for new resources, canceling services, changing existing resources, and requesting maintenance. This way of working tends to generate several problems:

1. It is difficult to know what was requested to the providers, when, and for whom. This happens when requests and their answers are restricted to the individuals who executed the request. There isn't usually a common depository where everything asked to the providers and everything answered by them is stored and made accessible to the telecom team.

2. It is difficult to coordinate parts of the processes that very often are executed by different groups within the telecom area. For example, one group may be responsible for requesting new resources; a second group may be responsible for controlling the provisioning, testing, and acceptance of resources; and a third group may be responsible for checking the invoices.

3. It is difficult to guarantee the quality of the requests if there is no standardized request form that guarantees that necessary information is sent to the providers properly and in a timely fashion (for example, complete address, site contact, contract, and details of the resource).

4. It is difficult to guarantee that the right communication channel is used; sometimes, a request is forwarded to the wrong people within the provider's staff. That may create a situation where the responsibility for problems becomes uncertain.

5. Registering requests (mostly maintenance requests) is done solely by the provider; in addition, each provider has its own preferred method of access (portal, e-mail, call center). In case of contention, the discussion will be based entirely on the provider's records (which isn't ideal from the organization's point of view).

6. It is difficult to link parts of the processes. For example, how can you guarantee that a resource is paid for only after it has been tested and accepted?

A good telecom resource management tool can help with all these points. Such a system can store all requests and provider answers in an organized way (it can even be the platform over which the requests themselves are made and answered). We usually define three institutional channels and require that all communications between the organization and its providers flow through them (one for the organization and two for the service providers—commercial and technical).

The process of interacting with the providers encompasses two phases:

- creating the requests
- following the requests

Interacting with service providers requires a flow of information in the following sequence:

1. The organization makes a request.
2. The provider answers whether it can or can't execute what was requested and, if yes, within which timeframe.
3. The organization approves the timeframe (or not; if not, the requests dies).
4. If the organization approves the request, the provider starts the execution and the time starts counting.
5. The provider finishes installing the resource and informs the organization.
6. The organization verifies whether the request was in fact executed and accepts the request as concluded (or not).

Maintenance requests go from phase 1 to 4 directly, since the time to execute is defined in contract.

Summarizing, it is imperative to adjust the internal procedures in order to implement a unified and standardized process of interacting with the service providers. The implementation of this process not only improves the quality of the interactions, it also reduces the time demanded to execute and manage them.

In addition, a unified process of requesting services reduces the dependence on certain individuals, putting the process in a more institutionalized level, where all information is shared with the team and where the communication between the organization and the provider is not personal.

Therefore, the crucial point is to establish institutional channels. A good telecom resource management system can be very helpful when it comes to implementing this process.

At this point, it is worth emphasizing that the idea of institutional channels is applicable in both ways; the organization sends requests through only one channel to only one recipient (or two, if we adopt the separation of commercial and technical channels), and the provider also answers back through only channel.

The provider may answer the request using the organization's software platform; this isn't common but should be tried wherever possible.

Although it may sound a bit obvious, when talking about institutional channels, we mean e-mails or phone numbers that are not associated with a specific person, such as an account manager.

This way of working through institutional channels, although sometimes difficult to implement, tends to increase productivity, simplifies the process, and makes it easier to document the interactions between the organizations. In addition, this strategy guarantees continuity independently of changes in the workforce of the organizations.

To implement institutional channels, we need to be able to negotiate with providers and implement internal mechanisms to forward the requests though a defined channel. Here, it is important to mention the fact that the establishment of defined institutional channels also helps enhance the security of the process, guaranteeing that only what was asked through the institutional channels is valid. A common problem is having multiple people asking for resources on behalf of the organization.

Chapter 7: Contract Management

It is absolutely crucial to have a defined process to control the contracts and tariffs. Aspects such as beginning and termination of the contracts and periodical adjustments in the tariff values are critical and must be very well controlled. It is important to have a process in place that alerts you when the contract's termination date is near (ideally at least three months before) and also when tariffs are going to undergo readjustments (annual value readjustment in pluri-annual contracts are common). This information must be linked with the processes of auditing the bills and renegotiating the contracts.

7.1 Unifying of Contracts

The first thing to do when organizing a telecom area is to unify the contracts as much as possible; it is very common in large organizations to see many more contracts than necessary. Having many contracts is counterproductive for several reasons:

- You have a multitude of bills with different cycles and payment days, which makes the payment and cost appropriation process much more difficult.
- You have a multitude of tariffs, which makes any process of negotiation, benchmarking, or bill auditing extremely complex.
- You have a multitude of renewal dates, which makes it extremely difficult to plan global renegotiation strategies.

Usually the process of unifying the contracts is made in parallel with resource inventory. After cleaning house, all resources can be consolidated into a new contract, and service providers can be told that anything not in the consolidated contract is canceled.

This strategy has its drawbacks, but it is usually more effective; if you had to wait for all your resources to be identified, in every business unit, before making the consolidation, you might never finish the process.

Another important factor associated with unifying contracts is to make all contracts terminate at the same time, standardizing the contract's lifespan. This will enable you to maximize the different services to be renegotiated (allowing wholesale negotiations).

7.2 Service Level Agreement of Telecommunications Contracts

Telecommunications contracts need to be managed and their QoS and SLA levels verified. This type of activity demands time and effort. It is a mistake to assume that the existence of control tools and rigid QoS clauses in the contracts, by themselves, will guarantee that failures are identified and penalties applied. The processes must be documented and also executed and audited to ensure effective execution.

We often see extremely rigid QoS clauses in contracts without a correspondent process in place to verify and punish the deviations. This situation makes an SLA (with additional associated costs) useless in the day-to-day running of the contract.

It is important to separate the SLA into two different types: the level of service associated with the quality of the telecom service provided and the quality of service associated with the requests made to the service providers.

- Quality of the telecom services: Key indicator factors are associated with network and resource availability (for example, monthly downtime below a certain level).

- Quality of services associated with the responsiveness of the providers: Key indicator factors associated with the responsiveness of the service provider are maximum time to respond to a request, maximum time to install a new resource, maximum time to cancel or change a resource, maximum time to fix a maintenance problem.

Therefore, in regard to the responsiveness of the service provider to the demands of the organization, there are at least five parameters to define in the contract:

1. **Time to answer a request:** Each time a request is made to the service provider, this request has to be answered within a defined number of hours. Note that "answer a request" doesn't mean actually doing what was required; it merely means acknowledging the demand and giving feedback about whether what was required is possible and within which time span.

2. **Time to install a new resource or cancel or change an existing one:** Of course, the time span to deliver a service may vary; however, a maximum time is usually agreed upon and should be enforced. In some cases, the provider determines the time span on a case-by-case basis.

3. **Time to accept (or not) the answer confirming the request:** This is the time given to the organization to confirm (or not) the request once the provider responds (for example, the organization requested a new last mile to a given site; the provider responded that it could install it within thirty days; the organization declines, saying that thirty days would be too long to wait, or the organization could also accept the time given).

4. **Time to test:** This is the time the organization has to test a resource, once the service provider informs it that the resource was installed, canceled, or changed.

5. **Time to have a maintenance problem solved:** This is the maximum time between the maintenance request and the complete solution of the problem.

Note that two of the parameters described apply to the organization, not to the service provider. This is natural, because this is a framework to define a relationship where the responsibilities of all parties are defined.

So it is advisable that all SLAs should be easily verifiable through proper tools and processes. From problem identification to the application of penalties, everything should be clearly defined, and all stakeholders should have their responsibilities very well defined.

A failure to actively manage the SLA components of a contract can lead to a situation where SLAs become mere guidelines for discussions (or justification for contract cancellation) when the relationship between a telecommunications provider and the organization has already deteriorated to a point where constructive interaction is not possible.

Chapter 8: Help Desk Management

The process of receiving a user's complaint about some malfunction in a telecom resource, registering this complaint, and forwarding it to a service provider to be solved is a very important part of managing a telecom infrastructure. This process encompasses two sets of activities: one internal and other external:

Internal: The internal part of the process encompasses all tasks involved with attending users' requests and registering these requests. Those are help desk processes. The execution of these processes requires some level of technical knowledge and the possibility of consulting the organization's knowledge base to identify known issues. Of course, this presupposes some sort of software program that documents requests, issues, and problems. Ideally, the organization already has an IT help desk to which requests regarding telecommunications resources can be forwarded. If there is not a structured IT help desk, it will be necessary to organize a specific telecommunications help desk, even if it has limited resources.

If the organization does not have some sort of organized process through which telecommunications problems are registered, the people of the telecommunications area will tend to spend a high percentage of their time attending requests directly from the users instead of solving those problems. This becomes more critical as the size of the organization grows.

Help desk operations may be outsourced, but the processes and documentation described here still apply. In fact, the documentation process becomes even more important when the people executing the tasks are not local or in-house.

A clear escalation procedure needs to be in place and communicated to the organization to catch any situation that may result in a customer satisfaction issue. Often, all the business needs is the assurance from senior management that the appropriate resources are deployed to address an issue.

External: Considering the fact that in most organizations, the bulk of the repair activity is executed by the service providers supplying the resources (trunks, links, mobile lines, etc.), it becomes absolutely critical to have some sort of mechanism in place to control the maintenance requests made by the organization to the service providers. Here we have to keep in mind some key points:

- All requests must be formal and numbered (as opposed to spoken requests made by phone).
- All requests must include a date, the time, and the name of the person making the request.
- All requests must include all details about the resource the complaint refers to.
- All requests must include a brief description about the problem as the user perceives it.

- All requests must be accessible for all the telecom team and not be an individual but rather a corporate instrument.

Here it is important to keep in mind that you must have service level agreements defining the time span to solve the problem with associated penalties in case the provider fails to do so.

Providers usually have call centers where problems can be reported. However, that doesn't eliminate the need for formal requests, for the following reasons:

- If you make your requests using only the provider's call-centers or tool, you will have a situation where the provider is the only one controlling the time spans associated with the requests (a clear conflict of interest). In case of any litigation about whether a request was solved on time or not, the only records will be from the service provider.

- If you make your requests using only the provider's call center or tool and you have many providers (which is common), you will have many processes associated with requesting maintenance. Having your own process gives you a unified view about your problems (throughout several providers) and a mechanism to follow them.

- Having your own process of generating maintenance requests to the provider forces some sort of unification of the requests by the organization, avoiding the problem of several groups within the organization requesting maintenance directly (sometimes the users themselves).

- Having your own process of generating maintenance requests to the provider allows you to standardize the information sent to the provider, guaranteeing that all data about the resources was correctly forwarded to the providers (which is very hard to guarantee in a verbal contact).

- Formalizing and unifying the maintenance requests allows control and enforcement of the level of services.

Chapter 9: Invoice Processing

Defining processes to control invoices is an important part of organizing the telecom control structure. Having a clear definition of how bills should be controlled is a crucial part of effectively controlling telecom costs.

Bills should be received in a central point (at least one central point per country). It is absolutely crucial to have a mechanism to signal if bills go missing. This is important because the organization must be able to detect if monthly bills that were supposed to be sent by the provider were not received.

Once you know that all bills were received, you should follow three basic steps:

1. **Register the value charged:** This is the value charged by the service provider, which is going to be confronted with the contractual values, historical values, and the results of the auditing verification.
2. **Register the value approved:** If there is a discrepancy between what is being charged and what the organization believes is the right value to be paid, you have to register that and indicate the discrepancy.
3. **Register the value effectively paid:** After the value charged and the value approved are discussed with the provider, you have to register how much was in fact paid.

The process of controlling these three values can take many forms, but you must be able to keep track of these values for each bill: How much I was charged, how much I believe I owe, and how much I actually paid.

These values should be controlled in a way that allows you to recover them using several criteria, such as by provider, by location, by contract, or by business unit (of course, indicating the value along with the time—typically monthly). This type of control allows you to follow historical values, which is usually an effective way to spot any problem with the charges.

Note the link between this process and the auditing bill process. Some organizations don't audit their bills every month; they only conduct an audit if they spot a discrepancy between the historical values and the value charged. We usually define four possible statuses for invoices (bills):

- **Waiting bill:** We know there is a bill, but the organization hasn't received it yet.
- **Waiting approval:** The organization received the bill but didn't verify if the value is due or not.
- **Waiting payment:** The organization got the bill, checked the correct value, and approved the payment, but the payment hasn't been made yet.
- **Paid:** The bill was paid.

These statuses are mutually exclusive (an invoice has only one status at a time), and they change according to the following sequence:

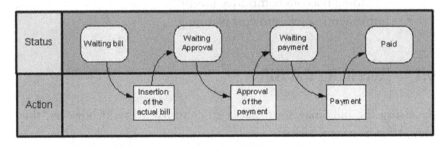

The following diagram gives us an example of how to organize the process to receive, verify, approve, and pay the bills.

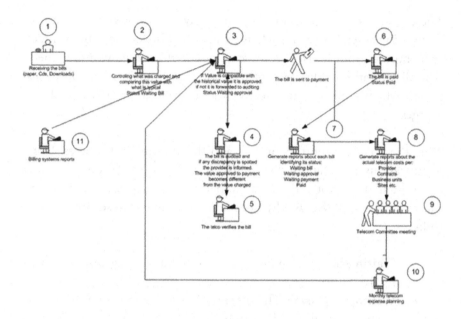

The picture above depicts how a bill is processed. The whole process encompasses eleven tasks, which could be grouped into six macro-tasks:

- receiving the bills
- checking the bill against the historical value
- auditing the bills and discussing discrepancies
- generating reports about the bills and costs
- attending telecom committee meetings
- implementing telecom cost planning

9.1   Receiving the Bills

Receiving the bills may seem like a straightforward task; however, they may arrive in several formats:

- Paper: The bill may be delivered by mail, having thousands of pages detailing the calls.
- Electronic media: The bill may be saved as an electronic file and provided on CD, delivered by mail.

- Electronic media: The bill may be made available on the provider's website to be downloaded by the client.

Bills may be delivered in different locations and in different levels of grouping. That said, it is essential to have a very well defined process to get all the bills from all providers. Of course, this does not only depend on the organization, but the ideal scenario would be to download the bill directly from the service provider's website. Even in this case, you need to define exactly when the bill will be available and who, when, and to where the download will be executed.

Simplifying this process is one of the reasons to unify the contracts as much as possible; grouping the bills makes the process of receiving them easier. For example, if you have an organization with two hundred sites and each one of them receives its own bill, putting all these bills together, not to mention processing them, will not be easy. This is why it is important to consolidate these bills into one and forward it to only one address.

Here you should also be able to notice if any bill is missing. This is a very common phenomenon, mostly when there are many bills and they are mailed to different addresses.

9.2   Checking the Bill against the Historical Value

After you receive the bill, the status of control changes from waiting for the bill to waiting for approval; that means the value is charged and you now must confirm that this value is compatible with the tariffs agreed and the resources charged are the ones contracted for.

At this point, you are also able to determine if a charge doesn't belong to the list of the bills that you pay every month. There are several possible verification strategies; they range from simply verifying the historical values to completely recalculating the bill. People often adopt a mixture of both strategies; for example, checking the compatibility of the value charged with the historical values of the bills with smaller values and entirely recalculating the big ones.

It is interesting to note that we usually have two types of procedures in regard to auditing the bills: you may do it for all bills every month and approve them only if the values are correct, or you may conduct an audit only when the value charged differs by a given percentage from what is typical.

In the previous picture, in procedures 2 and 3, the value charged is registered and then checked to ensure this value is compatible with the historical value. If it is, then the payment is approved. If it isn't, the bill is sent for verification (audited).

9.3 Auditing the Bills and Discussing Discrepancies

Although we will discuss the auditing of bills in more detail in the next chapter, it is important to link it with the bill processing. The input provided to the bill auditing process is the detailed bill and the value charged by the provider ("Value Charged," already registered by the group in charge of receiving the bills); the output of the bill auditing process will be the value considered to be correct ("Value Approved").

The discussion between the provider and the organization will determine the value that is in fact paid ("Value Paid"). Note that we have a situation here where the people who recalculate the bill (Step 4) may not be the same people discussing the discrepancies with the providers (Step 5), although if they are the same, the results tend to be better.

It is important to control all these values. Therefore, each bill has to have identified

- the valued charged,
- the value approved, and
- the value effectively paid (indicating the discrepancies between them).

In picture 14, that the person in charge of registering the value approved (step 3) is different from those in charge of auditing and discussing discrepancies with the service providers. This emphasizes the fact that they are distinct processes.

In regard to the process of disputing the values charged, there are typically three approaches:

1. Pay the bill as charged and require the reimbursement of the discrepancy in the next month.
2. Pay only the value considered correct and discuss the discrepancy later.
3. Don't pay the bill at all until the correct value is identified.

We believe that the second option is the fairest.

9.4   Generating Reports about the Bills and Costs

The telecom expense management group is the area responsible for recording all telecom resources used by the organization and identifying the associated costs. Therefore, it should be able to provide the following reports:

- list of providers
- list of sites
- list of contracts and tariffs
- list of resources contracted (per site, per contract, and per provider)
- expenditure per provider monthly
- expenditure per contract monthly
- expenditure per business unit monthly
- expenditure per site monthly
- expenditure per resource (last mile, equipment, and connection) monthly
- expenditure per cost center monthly
- list of the invoices created indicating value charged, value approved, and value actually paid
- detailed list of items of each invoice paid
- list of resources contracted, canceled, or changed

Here it is important to understand that the costs of the telecom infrastructure must be controlled, from the macro to the micro. Therefore, the first thing is to know how much the infrastructure costs as a whole and, from this total, drill down to the details. Of course, the process should ideally be constructed to provide the total costs and the details; however, in the real world, we may have to balance the effort to achieve granularity of information against the benefit of having the totalized information. For example, there is no point in trying to identify the types of calls of a specific site made (maximum granularity of the information), when you have sites where you don't even control all the bills. Telecom managers should achieve a uniform level of control over all sites and contracts of the organization; after the entire organization is covered, they can start drilling down the details. The drill-down should be done uniformly throughout the organization, always trying to guarantee that all sites and all contracts are controlled with the same level of granularity.

The information produced by the telecom expense management group usually feeds several groups within the organization:

- accounting
- financial and budgeting
- bill payments
- IT infrastructure
- operations

Each one of these groups usually needs to have this information formatted in different ways and within different time spans (monthly, weekly, quarterly, etc.). Here it is important to keep in mind that the telecom expense management group is in fact an enabler of other areas. It is the place within the organization where the bills are controlled and where all information regarding the inventory and the costs of all telecom resources are stored. Therefore, the telecom expense management area has two focuses: internal and external:

1. **Internal:** Control inventory, process bills, and control telecom costs
2. **External:** Provide the organization with all information regarding telecom inventory and costs

That said, the area has to be able to control the costs properly and provide the organization with precise and timely information. A telecom expense area that works looking only for his own tasks without concern in providing other areas with information is fulfilling only half of its responsibilities and tends to be like a bureaucratic branch, sterile regards possibilities of improvement and blind in regard to users' needs. We will discuss reports generated by the telecom expense area later.

9.5   Attending Telecom Committee Meetings

Telecom committee meetings are formal meetings conducted between the telecom people (including technical end cost control people) with the responsible for the user's areas. These meetings are usually held monthly and discuss the telecom costs of the previous month, trying to identify why the cost went up (or down) and determining the key performance indicators (KPIs) of the telecom structure. These meetings can also be used to discuss cost reduction strategies and to foresee variations in demand.

Making these meetings a regular and formal procedure is also very helpful in enhancing the relationship between users and the telecom area; the meetings also generate a level of control, since users will notice how much telecom cost represents and the telecom area will understand users' needs and problems better. Regular meetings between users and the control team forces the control team to attend to users' needs and avoids the bureaucratic attitude mentioned earlier.

Of course, these meetings only make sense when the costs are well controlled, to the point where the cause and effect of the ups and downs of the telecom cost can be traced.

9.6   Implementing Telecom Cost Planning

In large organizations, telecom has to be viewed as a basic resource necessary to make the organization operate, similar to energy, office space, and personnel. That said, it is important to budget telecom properly. Most companies identify the historical value spent on telecom and project this into the future. This methodology works if the environment remains stable, but it isn't effective if any kind of environmental variation happens. As we well know, variations happen all the time. Contract renewals, adding new services, choosing different calling strategies, and dealing with technical problems (just to mention some of the more frequent ones) can have a devastating effect on the telecom budget.

In order to make the planning process a bit less imprecise, we recommend including telecom planning as part of the telecom committee activities. In this context, users can state the services and strategies they are planning to implement, the technical people can describe the technical improvements or changes they are planning to make, and the telecom expense group can list the contractual and tariff changes they foresee. With these three views, it becomes possible to make a much better guess about how telecom costs will behave in the next budgeted period.

Planning the ROI of technical investments is in dissociable from the telecom budgeting process; this is pretty obvious but often overlooked.

It is very important to plan all activities impacting telecom costs. That said, we may have the following situations:

- The organization's call center may decide that it is going to increase the percentile of calls made to mobile phones because that increases the success rate of sales. It is a basic exercise to calculate how much more sales (and profits) are generated compared with the additional costs of making calls to mobile phones (which are usually more expensive than calls to fixed lines).
- The organization decides that it is going to change the PBX configuration policy regarding long distance calls, allowing a larger number of employees to make long distance calls.

These decisions can severely impact the telecom budget; sometimes, they aren't even worth the additional cost. The telecom committee is the forum where these initiatives should be discussed and the ROI calculated.

On the other hand, some projects and initiatives are aimed at reducing telecom costs; those initiatives should be reflected in the budget. The linkage between these initiatives and controlling present and future expenditures is crucial to fund those initiatives and, at the same time, to create a culture of foreseeing cost reductions and verifying if they are really happening.

Summarizing, we must try as hard as we can to have a clear view of all activities that will impact telecom costs, along with the time frame (usually one year ahead), identifying actions that will bring costs up, those that will push costs down, mapping how much each action represents in economic terms.

Of course, these recommendations only make sense if you already have a process in place where you can control your telecom costs.

9.7   Automatizing Bill Processing

When processing bills, people often view automatizing the process as a kind of panacea; downloading bills and automatically uploading them to a bill control application would be the final solution. Taking the risk to swim against the current, our view about this issue mirrors what Sir Alec Issigonis (designer of the Mini Cooper) said about comfort. He believed that drivers would be more alert if they were not sitting too comfortably, and so he deliberately made the seats a little uncomfortable.

Our view about bill processing is similar; the human eye is the ultimate cost control instrument, and any process that takes away this instrument tends to reduce the effectiveness of the control. That said, it is important to clarify that we are not against cost control tools; we just believe they must be a complement to human control, not a replacement for it. Bill processing should not run in automatic pilot mode.

Chapter 10: Auditing

We can define bill auditing as the process of verifying that what was charged by the service provider is what was agreed to in the contract. This verification includes calculating each call (in case of voice bills) and verifying every item charged in the bill.

Most organizations don't completely recalculate all bills every month; they usually do that only where there is a big discrepancy between what was identified by the billing system and what was actually charged, or when the value charged differs above a given percentage from the historical value.

Nevertheless, verifying bills is a necessity in a large organization. Verification identifies charge discrepancies and enables the organization to be reimbursed for overcharges. Our experience is that savings of 5 to 12 percent are attainable through regularly auditing the bills. Considering the telecom expenditures of large organizations, this percentage can represent an enormous amount of money; therefore, auditing the bills is usually worth the effort.

When we say "verify the bills," most people think about recalculating the value charged for the calls. Although this is an important part of the verification process, there are several other things that should be checked:

- Are the charges associated with trunks that actually belong to the organization?
- Are there excessive charges for installations or trunk subscriptions?
- Are there undue late payment penalties?
- Are the taxes calculated correctly?

- Are there charges for not achieving the minimum committed volume?

Are the charges associated with trunks that actually belong to the organization? Although it may seem a bit obvious, the first verification is to check if the bill belongs to the organization and the resource is actually being used; this basic review is often not done. It is not uncommon to receive bills that don't belong to the organization or that once did but the resource is now canceled or deactivated. In large operations with several sites (sometimes in different countries), verifying that the bills belong to the organization and the value charged is compatible with the historical value is in itself a difficult task.

Are there excessive charges for installations or trunk subscriptions? Eventual charges need to be carefully verified. This includes all sorts of charges, ranging from installations fees, trunk subscriptions, and special services. Those charges sometimes represent a large percentage of the bill, and their verification is far from easy (especially if you have several groups in the organization that request services from providers).

Are there undue late payment penalties? Penalties and interest due for late payments are often wrongly calculated or don't correspond with what was defined in the contract; the bill verifier must confirm that the penalties and interest are valid and calculated correctly (as defined in the contract).

This aspect of managing telecom costs is often overlooked; "other charges" include the following things:

- penalties for overdue payment
- interest for overdue payment
- charges for not complying with the minimum committed volume
- trunk subscription fees
- calls made through other service providers (common in those countries where you can choose a different service provider by dialing a particular prefix)

Here it is important to state the obvious: It is very important to pay the bills on time, and even if no recalculation of the bills is executed every month, some sort of control is necessary, even if only a simplified verification and comparison with the historical values. It is absolutely basic to control what invoices are paid, their due dates, and their typical values. If you don't receive the bill for any reason, it doesn't absolve you of late payment penalties.

Are the taxes calculated correctly? In some countries (and states, provinces, and cities within these countries), there are special tax breaks for specific services (call center operations, for example). This happens because call centers are usually very labor-intensive and governments may want to stimulate the creation of jobs. You may be entitled to discounts or rebates for your telecom bills. It is crucial to be aware of existing tax laws, which are often the defining reason for selecting the location of your sites. The auditing bill process often overlooks these possibilities.

Are there charges for not achieving the minimum committed volume? It is important to know if there are charges for not achieving the minimum committed volume and verify whether these charges were properly calculated (as defined in the contract). The difference between what was committed and what was actually used usually has to be paid in full.

These issues often lead to more overcharges than mistakes in the pricing of the calls.

Besides the direct financial gains, auditing the bills helps in many other important aspects linked with managing a large telecom infrastructure:

- identifying the delicate balance between the minimum volume commitment and tariffs
- identifying volumes and duration of the calls is crucial when negotiating charging granularity
- determining the point when least cost route configuration in the PBX needs adjustment due to changes in tariffs
- identifying when it becomes feasible to have a private voice network (such as when traffic is concentrated in a specific area code)
- accurately calculating the number of trunks and circuits (capacity planning)

Analyzing the voice bills provides accurate answers to all these questions; telecom managers should be aware that this analysis is much more than just checking the tariffs. It is also about verifying the traffic and comparing the current prices against available alternatives. The process of verifying the bill usually follows these steps:

- verify if the bill and the resources charged belong to the organization
- verify if the value charged is compatible with the historical value
- verify if the minimum committed volume was achieved (if not, recalculate the penalty and check if it was rightly charged)
- verify if there is any penalty and interest for late payment being charged
- verify if the charging period is correct and if there is any additional fee being charged for unsolicited services (installation, subscriptions, etc.)
- identify resources not in use (for example, trunks without any calls)

After verifying the basic items above, we should check the calls, recalculate the value, and check the traffic interest.

- verify from where and to where each call was made
- identify the area codes from and to each call
- verify how much each call is supposed to cost based on the organization's specific contract and identify the discrepancies
- identify if the taxes were properly applied
- verify if there are calls charged outside the admissible charging period

When auditing telephone bills, it is important to consider the fact that this procedure is effective only when there is a process in place to define how and when the organization will be reimbursed. Contracts must have dispute clauses, and ideally, the contracts in place should foresee that if errors were spotted in the bills before the payment due date, the organization can notify the service provider and pay only what was considered due. The values over which there is disagreement must be discussed jointly. If the charges prove to be right, the organization pays the service provider without penalties for delaying the payment (interest may be due).

The organization must define a formal process through which all invoice disputes are treated. The process needs a definition of time frames for each party involved and should mirror the dispute resolution clauses of individual contracts or master agreements. The whole process of disputing a bill from identification to solution should not exceed three months.

The SLA must reflect the fact that invoice payment doesn't imply an acceptance of the charged values. Ideally, the organization should have at least one year to audit the values charged. Ideally, the SLA should foresee that if identified errors exceed 5 percent of the total value of the bill, the service provider must reimburse the organization for the costs involved in auditing the bills; this is a considerable cost and a good mechanism for keeping service providers accurate.

The organization must define clearly the time after which charges are not acceptable (in some countries, it is defined by law); for example, services provided more than six months ago should not be charged.

The organization must define a regular schedule of meetings with the service provider invoice team. Such meetings are a forum to discuss problems with the invoices and items added, changed, and canceled.

The organization should try to define standardized invoice cycles so that all invoices are due on the same day; this simplifies the billing and payment process.

Of course, all these recommendations depend on negotiation at the contract stage, and these points arguably belong in chapter 2. However, these clauses only become meaningful if actual billing verification is done; without that, you will have no knowledge of billing errors or recourse when they are found.

Chapter 11: Billing Systems

The billing system is a very important component in controlling a telecom infrastructure. Through the billing system, it becomes possible to control the operation of the infrastructure, measuring traffic volumes and costs. Therefore, putting in place an effective and precise billing system is crucial. This tool will give you basic information about your traffic and enable you to follow your telco costs by type and spot discrepancies between what was actually charged and what was supposed to be charged. Basically, the main objective of a billing system is to guarantee an adequate level of control over telco costs. To achieve that, though, some considerations are necessary.

Initially, it is important to understand that large organizations usually deploy a billing system to control three types of environments:

- the organization's call centers
- the main administrative sites
- the operational sites

Each one of these environments has specific requirements, and the implementation of a billing system must take that into consideration.

The importance of the information generated also varies among these environments. Usually the Call centers are the ones where the need to control the calls and tariffs is more critical, followed by the main administrative sites. The data controlled by the billing system should be tailored to attend the specific needs of each of these environments.

Call Centers

- separation by campaign and services
- control of the incoming traffic
- route verification: identification of the route used as the first, second, or third alternative
- call classification
- totals by trunk groups

Large providers of call center services often use the billing system to separate traffic according to specific operations to allow this traffic to be charged back to the clients (clients are not necessarily charged exactly the same value charged by the telcos).

Main Administrative Sites

- billing per extension, department, and cost center
- billing by user through password
- reports indicating the outbound trunk group and traffic using internal voice channels (among the main sites)

Operational Sites

- billing per extension
- billing by user through password

You should guarantee that all trunks are connected to devices from where logs are collected and processed. If that isn't the case, you will have calls on your bill that are not verifiable by your billing system.

The process of collecting call detail records (CDRs) must be automatic (without human interference); otherwise, the process will not be as reliable as it should be.

The team in charge of the billing system must have unrestricted access to the telco contracts and tariffs; they must also keep the billing system database updated.

The team in charge of the billing system should also be in charge of defining the least cost routes and verifying the bills. This guarantees that the information related to tariffs, least cost routing (LCR), and bill totals per provider are thoroughly followed and the relations of cause and effect are clearly perceived. To successfully manage traffic and guarantee lower operational costs, it is crucial to have the same group of professionals seeing and understanding the costs (billing) and effectively acting over them (least cost routing).

A complete billing system usually encompasses devices to collect and store the call logs and servers to process and present these data. In large and heterogeneous environments, implementing an effective billing system is not a trivial task.

First, call log layouts should be standardized, because different equipment may generate logs with different formats; they should be normalized based on the format used by the billing software.

Second, logs of interconnected equipment should be processed without duplication. A clear understanding of the problem can be seen in the following example:

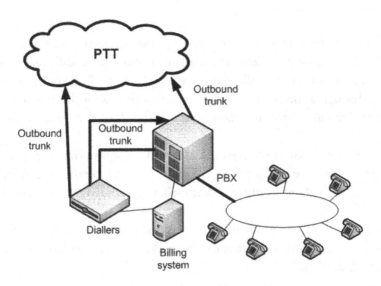

As we can see in this particular example, calls originating from dialers may go directly to the public network (generating a log only in the dialer), and calls originating from the dialer may go to a public network through the PBX (generating a log in the PBX and in the dialer). We should be able to separate these calls so that the same call is not identified by the billing system as two calls. This situation is even trickier if we can't just ignore the logs of one of the devices (we may need information about the extension from the PBX device and information about the mailing from the dialer device, or we may have calls generated directly by the PBX).

In addition to the difficulties associated with the initial installation, keeping the billing system working correctly demands a considerable effort to keep the databases updated with the right tariffs, trunks per equipment, and list of extensions.

Another important aspect is the need for cooperation between the installation teams and the team in charge of the billing system. It is very important to have a process in place that guarantees that any new equipment installed is linked to the billing system and that each time a trunk is installed or deactivated or any device is changed, this information is forwarded to the billing system team so the system database can be updated.

If there is no process in place to guarantee the connection of new devices to the billing system or detail trunk identification, some calls may not be collected or, if they are collected, they may be identified (and priced) as going through a specific telco when in reality they went through another. Being careful about this process guarantees that your data reflects reality.

It is also very important to separate the trunk groups in the billing system exactly as they are separated in the bills. It is important to generate a mirror of the bill and compare it with what you actually were charged.

It is critical to have a mechanism to keep the tariffs updated, and in case a tariff changes, it is important to have a clear identification of when the new tariff applies.

Ideally, a centralized billing system should collect and process the billing information of all sites and all devices. The system must be able to collect the billing data automatically; human intervention must be minimized.

The billing system must be able to identify in-/outgoing call parameters (including unidirectional and bidirectional trunks) and generate reports showing the following:

- number that originated or received the call (local, national, and international)
- number of the extension that originated or received the call
- date of the call (day/month/year)
- time the call started
- call duration
- user ID and original extension
- register of transfered calls

The billing system must be able to automatically and periodically (with the periodicity defined by the user) generate at least the following reports:

- all outbound calls (internal, local, national and international)
- all inbound calls (through direct access trunks)
- all calls by extension, indicating the external number dialed
- all calls whose duration exceeds a given time (in-and outbound)
- all calls executed through the operator consoles

Ideally, the system should be able to control the calls originated using user IDs, regardless of the extension that was actually used. In addition, the system should be able to control calls transferred, generating a specific log for each transfer.

In terms of guarantee operational continuity, the billing system (here including the whole apparatus) must be able to insure that the billing process continues in the case of minor failures. Therefore, the structure should be designed to store the call logs for some reasonable period of time (if the connection with the billing workstation becomes unavailable).

The designer of the structure must guarantee that all devices being billed are equipped with internal storage capacity to support at least eight hours (typical) of operation without the need for unloading the logs to the billing storage system.

For the most important devices, you may consider using external buffers to support the storage of the logs for an additional number of hours.

The system must have mechanisms to identify clearly if some device failed to provide logs during some period of time.

When defining the requirements of the system, you should guarantee that it complies with the following:

- be adaptable to different configurations and needs
- have an acceptable performance during the highest traffic hour
- be flexible to comply with changes and upgrades

You may also specify the operating system (OS) and database to be used by the billing software itself.

In large installations, it is advisable to divide the billing system functionalities into two independent modules: data collection/storage and management. This is a good practice not only because you can separate the functionalities into two devices and use the redundancy as a backup (one device may be able to execute the other's functionality), but also because you may be able to operate the management system without loss of performance during log collection and processing.

The data collection process must enable you to receive and store all logs generated by all network modules. The management system in its turn must be able to treat the data collected, providing reports, the inclusion or exclusion of data, and configuration parameters (e.g., tariffs).

You should make sure that the billing system is able to collect the logs from the devices using different strategies: direct connection, connection through a LAN, or dial-in. The data collection must happen automatically and independent to the model of the equipment in each site. The provider of the billing system must be able to adjust all the necessary interfaces and make all the necessary adjustments to the file layout.

If you have sites located in different countries, the billing system must be able to treat more than one currency, converting and consolidating costs.

When procuring a billing system, you should provide potential providers with a general view of the volumes involved and the number of calls generated by month and by day. Based on this information, the provider can specify the hardware capacity required and the maximum number of calls that the system can handle (spare capacity is advisable). Providers can state the limits of the proposed solution and explain any expansion alternatives. Remember, it is your responsibility to tell them your volume; they don't know that.

You should emphasize to potential providers that their proposal should include all devices and software associated with transporting the logs from the devices to the billing workstation(s). For example, if you require the logs to be collected through the company LAN and a new router is required, it should be included in the proposal (all additional hardware and software necessary to the implementation/operation of the billing system as a whole).

Remote access is a very useful feature that is often overlooked by telecom managers when preparing the technical specifications of a billing system. Ideally, the billing system must be remotely configurable (including all parameters). This feature must include all sites as well as the main unit. Remote access must allow at least the following features:

- set the date and hour
- switch specific parameters
- set the data transference time
- access through dial-in
- verification of the level of storage and memory use
- set the data collection mode (on demand or automatic)
- adjustments in the data format conversion (text to the database used by the billing system)
- access to some billing reports

When defining the requirements of a billing system, it is important to make sure it can be accessed through a network; you should be able to expand it using as many workstations the increase of volume may demand. It is good to know the expansion cost in advance; in other words, the system must be scalable in all its features and you should know how much it costs to expand it.

All systems must have password-controlled access, and users may be divided into different levels (preferably defining user access individually). In addition, the system should be user friendly to make training easier, and all reports should be visible on the screen before being printed using the standard WYSIWYG (What You See Is What You Get).

The ideal system should have a report generator with filters to select the calls that match a particular set of requirements (data, cost, duration, extension, etc.). In addition, it is ideal to be able to save the report formats in some sort of "gallery."

The report generator should be able to generate reports of calls by at least the following criteria (detailed or grouped):

- costs centers
- extensions
- trunks and tie-lines
- special tariffs
- number dialed
- external telephone list
- type of expense codes
- additional fees
- tariffs
- users
- operator's desk
- tariff per service provider
- duration or cost above a given value
- forwarding prescheduled reports automatically via e-mail

Another interesting feature to have is storing a telephone number and generating a report indicating how many times this particular number was called (or how many calls from it were received).

The system must be able to make security backups automatically through the network and allow database exportation for different formats, such as XLS, PDF, DBF, or TXT, allowing easy integration with external systems—Open database.

Another important functionality is the simulation and recalculation of logs based on a different set of tariffs.

Ideally, the system should provide an audit trail, indicating all interventions and processes operated by the system. That means indicating things such as what each user did or what treatment each billing log or call detail report required.

11.1   Effectiveness of the Billing System

The effectiveness of a billing system very often has more to do with the way it is installed and operated than with the software itself. That said, we will now analyze the main aspects that make controlling bills most effective:

11.1.1   Aspects Linked with the Installation

Depending on how the hardware is connected, you may have problems collecting all logs automatically and have to add part of the logs manually (even if only temporarily); in addition, if you don't have very thorough control and a device fails to send its logs, you may end up generating reports that don't reflect what really happened.

In situations when devices are interconnected, you may have a situation where the same call generates logs in more than one device. This has to be very carefully controlled. Here is worth mentioning the fact that in some circumstances you cannot just ignore the logs of one of the devices, because these logs may have relevant information (for example, one device generates the call and knows the extension used, and another forwarded the call to the public network and knows which provider was used).

Another common problem is if the system incorrectly identifies the trunk that made a call. If trunk identification isn't updated, the billing system will identify the calls originated through a specific trunk as being supported by one service provider, when in reality it was supported by other. That means the cost of the call is wrongly calculated. This is a common problem and occurs due to miscommunication between the installation teams and the billing team (a failure to communicate moves and changes to the billing team).

Very often, trunks are not grouped exactly as the service providers charge them. This prevents the organization from generating a truly mirrored bill, which in turn jeopardizes cost control.

11.1.2   Aspects Linked with the Process

Very often, the team in charge of the billing process is not associated with those in charge of negotiations and contract management. In this situation, new tariffs or billing rules may be negotiated without informing the billing team.

In the same way, we may find a situation where the billing team isn't linked with the call-routing team, which makes it harder to effectively manage call routing. This sounds strange, but it is a very common scenario.

Chapter 12: Traffic Analysis and Optimization

To be able to effectively manage the telecom resources in any large organization, it is crucial to have a clear understanding of the traffic being supported by the telecom structure. It is very important to know the volumes per type of service, the interest (from where to where the flows go), the profile (how the flows behave along the time), and of course the costs associated with transporting the flows.

Therefore, it is absolutely necessary to check the sources and destinations of the traffic, identifying from where and to where the flows go ("flow" here means voice and data traffic). Close control of this information makes it possible to identify strategies to reduce costs and negotiate properly with the service providers. Traffic must be controlled according to the following aspects:

- traffic volume (quantity of calls and connections: kilobits, packets, or minutes)
- traffic interest (from where to where: area codes, countries, states, cities)
- typical hourly distribution, including hourly concentration factor (HCF)
- typical daily distribution, including daily concentration factor (DCF)
- interconnection costs (cost to transport the flows, usually tariffs)
- volume per site and service
- traffic by type (for example, outbound to mobile, inbound from mobile, Internet traffic)

- traffic by service provider (originated and terminated)
- duration patterns (typical duration of calls, file transferences, data sections, etc.)
- routing traffic (least cost routing)
- traffic concentration (quantity of originating and receiving ends)

Therefore, identification of current usage patterns is critical for all planning initiatives. Not knowing how your traffic behaves makes it hard to negotiate things like tariffs, charging granularities, and minimum committed volumes.

When identifying opportunities for cost reductions, telecom managers in a large organization must consider several factors, but in general, these factors can be narrowed down to two main lines of action (not mutually exclusive):

- negotiating good prices and charging strategies
- routing the flows properly

Most issues are linked to knowing the types, profiles, and durations of the flows (voice and data) and using this information to get the best tariffs as possible.

For example, in the case of voice traffic, knowing the typical duration of calls, from where they are originated, and to where they go helps to identify which charging granularity is more suitable and also helps to negotiate special prices for specific locations.

Charging Granularity

This is the minimum amount of time charged per connection (voice calls or nondedicated data connections). For example, a charging granularity of 30s+6s+6s for voice calls indicates that if a call is established, it is charged as if its duration was thirty seconds, even if it was less. After thirty seconds, the service provider charges an entire quantum of six seconds each time span below or equal to six seconds. Therefore, using this charging granularity as an example, a call lasting eighteen seconds would be charged as thirty seconds. In the same way, a call lasting thirty-one seconds would be charged as thirty-six seconds (thirty seconds plus six seconds); a call lasting thirty-eight seconds would be charged as forty-two seconds (thirty seconds plus six seconds, plus six seconds); and so forth.

Therefore, if you have a large percentage of calls lasting less than thirty seconds (short calls), which would be very typical if you have an active call center, it will make a huge difference if you have a charging granularity of 6s+6s+6s instead 30s+6s+6s, for example. The same concept applies to data services.

Each case has its own particularities; however, the concept has to be well understood and the traffic profile well mapped before conducting a negotiation. What really counts is the average cost of the unit charged (minute, kilobit, packet, etc.) and the balancing act between charging granularity/price per unit and average volume of the transported units.

Knowing from where to where the flows come and go also makes a huge difference when negotiating the tariffs. This allows you to get better prices in those areas where your traffic is bigger. This is pretty obvious but very often overlooked.

Another important point when negotiating large contracts is what is called "minimum committed volume." Telco contracts usually demand a minimum monthly payment corresponding to a minimum volume, whether actually used or not. For example, if you have a voice contract stating that the minimum committed volume is USD 100,000 per month and your cost per minute is USD 0.01, you have to pay at least $100,000 every month even if that amount of minutes was not used. Usually, these negotiations are delicate balancing acts, given the fact that the cost of the minutes usually goes down as the minimum committed volume goes up. Knowing the average volume of the organization and the volume expected to be forwarded through a specific telco, given a specific tariff, is crucial in a negotiation process.

Here is a strategic consideration: Even when a specific telco has the best prices and conditions, it is wise to not rely on only one provider for all your traffic. The ideal scenario is to divide the traffic between at least two providers. This is important for both technical and commercial reasons.

Technical reasons: Any provider may experience technical outages, and it is crucial to have an alternative or at least the possibility of partially continuing operations (backup circuits, for example, must always be contracted from different providers).

Commercial reasons: It is always good to have a ready alternative if some commercial problem arises. Providers should know that there are alternative sources you can easily shift your traffic to if you need or want to (with the resources already installed).

For these reasons, beware of the mermaid's song, when an existing provider comes to you with a potential discount if you raise your minimum committed volume. You may have to redirect traffic from other providers, relying too much on only one. You may also default on the minimum committed volume of the other contracts and be forced to pay minimum values for them. The discount may not be worth it.

The ideal situation is having at least two contracts, each one with a minimum committed volume not exceeding 25 percent of your actual volume. This situation gives you some leeway to shift traffic between the two providers without incurring penalties. Of course, everything depends on the volume discount and how critical your operation is.

It is important be able to redirect traffic. We will discuss this aspect in more detail later; however, it is important to keep in mind that being able to redirect the traffic flow makes a huge difference in negotiating with service providers. If the service provider knows you don't have quick alternatives in the short term, it will be tempted to play hardball.

12.1   Traffic Volumes

It is very important to understand the type of traffic your organization has and the objectives associated with the generation of flow and volume. We also have to understand the seasonality of the traffic and how it varies along the months of the year. The telecom manager has to have a clear view of the volumes, considering several aspects:

By interest of the flows (inbound and outbound):

Type of traffic	Minutes	%
Inbound	115,107.50	1.29%
Outbound	8,831,515.61	98.71%
Total	8,946,623.11	100.00%

By type (voice mobile local, voice fixed long distance, international, etc.):

Type of traffic	Minutes	%
Fix to fix local	2,591,501.09	28.97%
Fix to fix long distance	5,888,906.92	65.82%
Fix to mobile local	252,343.34	2.82%
Fix to mobile long distance	213,871.76	2.39%
Total	8,946,623.11	100.00%

In addition, it is necessary to know how many providers you have and the percentage of the volume (per type) each one of them supports:

Provider	Type of service	Minutes	Value	% Minutes	% Values
Telco 1	Fix to fix local	1.866.772,50	USD 78.663,92	20,87%	7,38%
	Fix to fix long distance	4.371.791,00	USD 388.756,16	48,87%	36,46%
	Fix to mobile local	176.224,10	USD 183.385,46	1,97%	17,20%
	Fix to mobile long distance	85.479,40	USD 65.560,80	0,96%	6,15%
Telco 2	Inbound fix to fix local	34.917,00	USD 7.573,41	0,39%	0,71%
	Inbound fix to fix long distance	79.353,40	USD 20.014,90	0,89%	1,88%
	Inbound fix to mobile local	458,90	USD 34,20	0,01%	0,00%
	Inbound fix to mobile long distance	378,20	USD 29,01	0,00%	0,00%
	Outbound fix to fix local	36.124,30	USD 961,61	0,40%	0,09%
	Outbound fix to fix long distance	144.466,40	USD 14.431,86	1,61%	1,35%
Telco 3	Fix to fix local	101.444,60	USD 15.087,67	1,13%	1,41%
Telco 4	Fix to mobile local	23.528,40	USD 9.246,23	0,26%	0,87%
	Fix to mobile long distance	102.727,00	USD 71.100,43	1,15%	6,67%
Telco 5	Fix to fix local	552.242,69	USD 23.216,65	6,17%	2,18%
	Fix to fix long distance	1.293.296,12	USD 114.736,42	14,46%	10,76%
	Fix to mobile local	52.131,94	USD 54.123,88	0,58%	5,08%
	Fix to mobile long distance	25.287,16	USD 19.349,43	0,28%	1,81%
Total		8.946.623,11	USD 1.066.272,04	100,00%	100,00%

It is also important to know the traffic as a whole, separating the part paid directly by your organization, the part paid directly by clients/users, and the part paid by your organization and later reimbursed by clients/users.

12.2   Traffic Interest

We have to identify the voice and data flows by location in terms of number of connections, duration of these connections, and value. This information allows us to understand the traffic behavior and gives us an instrument to act properly. The following traffic matrixes show how this information should be presented.

The data traffic matrixes represent the number of circuits, bandwidth, and cost to provide data connectivity between the organization's sites. For the sake of space, we are demonstrating the traffic matrix of an organization with only nine sites, but the idea remains the same for a larger organization:

Number of circuitos

Site number	Address	010	001	022	002	070	003	032	004	015	005	006	007	008	009
001	BRAA AE 4 LT 1 ST TRADICIONAL	1	0	0	0	0	0	0	0	0	0	0	0	0	0
002	BRAA AE 4 LT 3 ST TRADICIONAL	0	0	1	0	0	0	0	0	0	0	0	0	0	0
003	BRAA AV VEREDINHA AE SN ST TRADICIONAL	0	0	0	0	1	0	0	0	0	0	0	0	0	0
004	BRAA CCD (CENTRO COM DIVERSOES) BL 1 LJ 4 ST NORTE	0	0	0	0	0	0	1	0	1	0	0	0	0	0
005	BRAA Q 15 INCRA 8 LT 2 ALEXANDRE DE GUSMAO	1	0	0	0	0	0	0	0	0	0	0	0	0	0
006	BSA CRS 507 BL C LJ 41/69	1	0	0	0	0	0	0	0	2	0	0	0	0	0
007	BSA ETC RODOVIARIA	0	0	1	0	0	0	0	0	4	0	0	0	0	0
008	BSA ETD MANE GARRINCHA AE SN	2	0	2	0	0	0	0	0	2	0	0	0	0	1
010	BSA PC BURITI ANEXO PALACIO BURITI	4	0	6	0	0	1	0	0	0	0	0	0	0	0
015	BSA SAIN BL H SL CPD CODEPLAN	0	2	0	0	3	0	0	0	3	0	0	0	0	0
022	BSA SBN Q 2 BL K AN DO ED WAGNER	2	0	1	0	1	0	0	0	0	0	0	0	0	0
032	BSA SCS Q 8 BL B60 S 240 VENANCIO 2000	1	0	0	0	0	0	0	0	2	0	0	0	0	0
070	NBDE ST ND BERNARDO SAYAO Q 1 CJ B LT 14	0	0	0	0	0	0	0	0	0	0	0	0	0	0

Bandwidth

Site number	Address	010	001	022	002	070	003	032	004	015	005	006	007	008	009
001	BRAA AE 4 LT 1 ST TRADICIONAL	256	0	0	0	0	0	0	0	0	0	0	0	0	0
002	BRAA AE 4 LT 3 ST TRADICIONAL	0	0	64	0	0	0	0	0	0	0	0	0	0	0
003	BRAA AV VEREDINHA AE SN ST TRADICIONAL	0	0	0	0	256	0	0	0	0	0	0	0	0	0
004	BRAA CCD (CENTRO COM DIVERSOES) BL 1 LJ 4 ST NORTE	0	0	0	0	0	0	32	0	64	0	0	0	0	0
005	BRAA Q 15 INCRA 8 LT 2 ALEXANDRE DE GUSMAO	256	0	0	0	0	0	0	0	0	0	0	0	0	0
006	BSA CRS 507 BL C LJ 41/69	512	0	0	0	0	0	0	0	128	0	0	0	0	0
007	BSA ETC RODOVIARIA	0	0	512	0	0	0	0	0	1024	0	0	0	0	0
008	BSA ETD MANE GARRINCHA AE SN	1024	0	1024	0	0	0	0	0	128	0	0	0	0	512
009	BSA GN WILSON NELSON BL 1	256	0	0	0	0	0	0	0	0	0	0	0	0	3
010	BSA PC BURITI ANEXO PALACIO BURITI	2432	0	3200	0	0	256	0	0	1536	0	0	0	0	0
015	BSA SAIN BL H SL CPD CODEPLAN	0	120	0	0	1040	0	0	0	0	0	0	0	0	0
022	BSA SBN Q 2 BL K AN DO ED WAGNER	384	0	128	0	1024	0	0	0	0	0	0	0	0	0
032	BSA SCS Q 8 BL B60 S 240 VENANCIO 2000	1024	0	0	0	0	0	0	0	128	0	0	0	0	0
070	NBDE ST ND BERNARDO SAYAO Q 1 CJ B LT 14	0	0	0	0	0	0	0	0	0	0	0	0	0	0

Cost

Site number	Address	010	001	022	002	070	003	032	004	015	005	006	007	008	009
001	BRAA AE 4 LT 1 ST TRADICIONAL	USD 999,00	USD 0,00	USD 0,00	USD 0,00	USD 0,00	USD 0,00	USD 0,00	USD 0,00	USD 0,00	USD 0,00	USD 0,00	USD 0,00	USD 0,00	USD 0,00
002	BRAA AE 4 LT 3 ST TRADICIONAL	USD 0,00	USD 0,00	USD 385,00	USD 0,00	USD 0,00	USD 0,00	USD 0,00	USD 0,00	USD 0,00	USD 0,00	USD 0,00	USD 0,00	USD 0,00	USD 0,00
003	BRAA AV VEREDINHA AE SN ST TRADICIONAL	USD 0,00	USD 0,00	USD 0,00	USD 0,00	USD 999,00	USD 0,00	USD 0,00	USD 0,00	USD 0,00	USD 0,00	USD 0,00	USD 0,00	USD 0,00	USD 0,00
004	BRAA CCD (CENTRO COM DIVERSOES) BL 1 LJ 4 ST NORTE	USD 0,00	USD 0,00	USD 0,00	USD 0,00	USD 0,00	USD 0,00	USD 278,00	USD 0,00	USD 385,00	USD 0,00	USD 0,00	USD 0,00	USD 0,00	USD 0,00
005	BRAA Q 15 INCRA 8 LT 2 ALEXANDRE DE GUSMAO	USD 999,00	USD 0,00	USD 0,00	USD 0,00	USD 0,00	USD 0,00	USD 0,00	USD 0,00	USD 0,00	USD 0,00	USD 0,00	USD 0,00	USD 0,00	USD 0,00
006	BSA CRS 507 BL C LJ 41/69	USD 2.422,00	USD 0,00	USD 0,00	USD 0,00	USD 0,00	USD 0,00	USD 0,00	USD 0,00	USD 771,00	USD 0,00	USD 0,00	USD 0,00	USD 0,00	USD 0,00
007	BSA ETC RODOVIARIA	USD 0,00	USD 0,00	USD 2.422,00	USD 0,00	USD 0,00	USD 0,00	USD 0,00	USD 0,00	USD 3.396,00	USD 0,00	USD 0,00	USD 0,00	USD 0,00	USD 0,00
008	BSA ETD MANE GARRINCHA AE SN	USD 4.844,00	USD 0,00	USD 4.844,00	USD 0,00	USD 0,00	USD 0,00	USD 0,00	USD 0,00	USD 771,00	USD 0,00	USD 0,00	USD 0,00	USD 0,00	USD 2.422,00
009	BSA GN WILSON NELSON BL 1	USD 999,00	USD 0,00	USD 0,00	USD 0,00	USD 0,00	USD 0,00	USD 0,00	USD 0,00	USD 0,00	USD 0,00	USD 0,00	USD 0,00	USD 0,00	USD 0,00
010	BSA PC BURITI ANEXO PALACIO BURITI	USD 10.844,00	USD 0,00	USD 14.265,00	USD 0,00	USD 0,00	USD 999,00	USD 0,00	USD 0,00	USD 6.356,00	USD 0,00	USD 0,00	USD 0,00	USD 0,00	USD 0,00
015	BSA SAIN BL H SL CPD CODEPLAN	USD 0,00	USD 771,00	USD 0,00	USD 0,00	USD 4.969,00	USD 0,00	USD 0,00	USD 0,00	USD 0,00	USD 0,00	USD 0,00	USD 0,00	USD 0,00	USD 0,00
022	BSA SBN Q 2 BL K AN DO ED WAGNER	USD 1.699,00	USD 0,00	USD 999,00	USD 0,00	USD 4.039,00	USD 0,00	USD 0,00	USD 0,00	USD 0,00	USD 0,00	USD 0,00	USD 0,00	USD 0,00	USD 0,00
032	BSA SCS Q 8 BL B60 S 240 VENANCIO 2000	USD 4.826,00	USD 0,00	USD 0,00	USD 0,00	USD 0,00	USD 0,00	USD 0,00	USD 0,00	USD 771,00	USD 0,00	USD 0,00	USD 0,00	USD 0,00	USD 0,00
070	NBDE ST ND BERNARDO SAYAO Q 1 CJ B LT 14	USD 0,00	USD 0,00	USD 0,00	USD 0,00	USD 0,00	USD 0,00	USD 0,00	USD 0,00	USD 0,00	USD 0,00	USD 0,00	USD 0,00	USD 0,00	USD 0,00

The voice traffic matrixes usually demonstrate the voice traffic between area codes (or if we are mapping the internal voice traffic, we may identify the traffic among the organization's sites, as in the data traffic matrix). Usually we represent the voice traffic matrixes per number of calls, minutes, and cost:

Number of calls

Area Codes	Area codes names	477	311	442	461	464	55	618	656	993	984	444	614	639	771
477	LEON, GTO	3	10	221	24	14	331	26	20	0	3	62	0	0	0
311	TEPIC, NAV	0	0	0	0	0	0	0	0	0	0	0	0	0	5
442	QUERETARO, QRO	63	0	3	38	30	724	0	0	1	0	2	0	0	195
461	CELAYA, GTO	0	0	0	0	0	0	0	0	0	0	0	0	0	5
464	SALAMANCA, GTO	0	0	0	0	0	0	0	0	2	0	0	0	0	3
55	MEXICO Y AREA METROPOLITANA, DF	6005	693	7206	1356	536	923	1867	7088	3801	1385	6893	4815	459	3147
618	DURANGO, DGO	0	0	0	0	0	114	4	56	0	0	0	227	0	0
656	CIUDAD JUAREZ, CHIH	0	0	1	0	0	452	20	2	0	0	1	203	1	0
993	VILLAHERMOSA, TAB	0	0	4	0	1	163	0	0	0	5	0	0	0	3
984	AKUMAL, QROO	0	0	0	0	0	169	0	0	2	1	0	0	0	0
444	SAN LUIS POTOSI, SLP	120	0	8	0	0	71	0	0	0	0	4	0	0	3
614	CHIHUAHUA, CHIH	0	3	2	0	0	573	41	390	9	0	0	4	535	0
639	CIUDAD DELICIAS, CHIH	5	0	0	0	0	3	0	2	0	0	0	0	0	0
771	PACHUCA, HGO	13	0	809	0	1	298	0	0	0	0	6	0	0	1

Minutes

Area codes	Area code name	477	311	442	461	464	55	618	656	993	984	444	614	639	771
477	LEON, GTO	3	18	687	59	27	966	92	56	0	18	199	0	0	13
311	TEPIC, NAV	0	0	0	0	0	0	0	0	0	0	0	0	0	0
442	QUERETARO, QRO	362	0	3	118	116	3299	0	0	77	0	15	0	0	829
461	CELAYA, GTO	0	0	0	0	0	0	0	0	0	0	0	0	0	2
464	SALAMANCA, GTO	0	0	0	0	0	0	0	0	0	0	0	0	0	5
55	MEXICO Y AREA METROPOLITANA, DF	12291	1837	18295	2752	1093	1748	3908	14890	12875	3378	13296	9556	671	8760
618	DURANGO, DGO	0	0	0	0	0	461	4	184	0	0	0	1053	0	0
656	CIUDAD JUAREZ, CHIH	0	0	17	0	0	1574	30	2	0	0	2	918	0	0
993	VILLAHERMOSA, TAB	0	0	43	0	3	498	0	0	0	63	0	0	0	15
984	AKUMAL, QROO	0	0	0	0	0	394	0	0	63	1	0	0	0	1
444	SAN LUIS POTOSI, SLP	490	0	12	0	0	246	0	0	0	0	4	0	0	79
614	CHIHUAHUA, CHIH	0	74	3	0	0	1194	260	1268	10	0	0	4	1068	0
639	CIUDAD DELICIAS, CHIH	0	0	0	0	0	6	0	3	0	0	0	0	0	0
771	PACHUCA, HGO	131	0	2691	0	1	796	0	0	0	0	10	0	0	1

Cost

Area Code	Area Code Name	477	311	442	461	464	55	618	656	991	904	994	614	639	771
477	LEON, GTO	USD 0,00	USD 1,00	USD 50,00	USD 4,00	USD 2,00	USD 78,00	USD 7,00	USD 4,00	USD 0,00	USD 1,00	USD 44,00	USD 0,00	USD 0,00	USD 1,00
311	TEPIC, NAV	USD 0,00	USD 0,00	USD 0,00	USD 0,00	USD 0,00	USD 0,00	USD 0,00	USD 0,00	USD 0,00	USD 0,00	USD 0,00	USD 0,00	USD 0,00	USD 0,00
442	QUERETARO, QRO	USD 24,00	USD 0,00	USD 0,00	USD 0,00	USD 10,00	USD 227,00	USD 0,00	USD 0,00	USD 5,00	USD 0,00	USD 1,00	USD 0,00	USD 0,00	USD 55,00
461	CELAYA, GTO	USD 0,00	USD 0,00	USD 0,00	USD 0,00	USD 0,00	USD 0,00	USD 0,00	USD 0,00	USD 0,00	USD 0,00	USD 0,00	USD 0,00	USD 0,00	USD 0,00
464	SALAMANCA, GTO	USD 0,00	USD 0,00	USD 0,00	USD 0,00	USD 0,00	USD 0,00	USD 0,00	USD 0,00	USD 0,00	USD 0,00	USD 0,00	USD 0,00	USD 0,00	USD 0,00
55	MEXICO Y AREA METROPOLITANA, DF	USD 800,00	USD 120,00	USD 1.196,00	USD 182,00	USD 71,00	USD 136,00	USD 228,00	USD 949,00	USD 839,00	USD 221,00	USD 869,00	USD 632,00	USD 57,00	USD 447,00
618	DURANGO, DGO	USD 0,00	USD 0,00	USD 0,00	USD 0,00	USD 0,00	USD 31,00	USD 1,00	USD 12,00	USD 0,00	USD 0,00	USD 0,00	USD 70,00	USD 0,00	USD 0,00
656	CIUDAD JUAREZ, CHIH	USD 0,00	USD 0,00	USD 0,00	USD 0,00	USD 0,00	USD 105,00	USD 2,00	USD 0,00	USD 0,00	USD 0,00	USD 0,00	USD 61,00	USD 0,00	USD 0,00
993	VILLAHERMOSA, TAB	USD 0,00	USD 0,00	USD 3,00	USD 0,00	USD 0,00	USD 33,00	USD 0,00	USD 0,00	USD 0,00	USD 3,00	USD 0,00	USD 0,00	USD 0,00	USD 1,00
984	AKUMAL, QROO	USD 0,00	USD 0,00	USD 0,00	USD 0,00	USD 0,00	USD 26,00	USD 0,00	USD 0,00	USD 4,00	USD 0,00	USD 0,00	USD 0,00	USD 0,00	USD 0,00
444	SAN LUIS POTOSI, SLP	USD 33,00	USD 0,00	USD 1,00	USD 0,00	USD 0,00	USD 16,00	USD 0,00	USD 0,00	USD 0,00	USD 0,00	USD 1,00	USD 0,00	USD 0,00	USD 5,00
614	CHIHUAHUA, CHIH	USD 0,00	USD 5,00	USD 0,00	USD 0,00	USD 0,00	USD 80,00	USD 18,00	USD 84,00	USD 1,00	USD 0,00	USD 0,00	USD 1,00	USD 81,00	USD 0,00
639	CIUDAD DELICIAS, CHIH	USD 0,00	USD 0,00	USD 0,00	USD 0,00	USD 0,00	USD 0,00	USD 0,00	USD 0,00	USD 0,00	USD 0,00	USD 0,00	USD 0,00	USD 0,00	USD 0,00
771	PACHUCA, HGO	USD 9,00	USD 0,00	USD 182,00	USD 0,00	USD 0,00	USD 53,00	USD 0,00	USD 0,00	USD 0,00	USD 0,00	USD 1,00	USD 0,00	USD 0,00	USD 0,00

In this particular example, the traffic among only five area codes and DF/ Mexico City (55) represents 25 percent of the whole cost. This fact sets the direction for an eventual negotiation and opens the door to consider the possibility of building a private voice network interconnecting these five locations.

To have a better view of the traffic, we usually divide the traffic by state of destination/origin; it may help if you have telcos that operate predominately in a given state or area. The following spreadsheet shows how a Brazilian organization's traffic is distributed by state:

State	Quant	Value	Minutes	% Quant	% Value	% Minutes
SP	5,278,500.43	USD 522,079.64	3,939,179.43	46.71%	48.96%	44.03%
RJ	1,074,432.80	USD 107,127.01	976,941.49	9.51%	10.05%	10.92%
MG	880,159.37	USD 80,686.14	741,237.78	7.79%	7.57%	8.29%
BA	478,340.65	USD 43,718.82	402,119.44	4.23%	4.10%	4.49%
RS	534,914.29	USD 42,904.81	399,383.63	4.73%	4.02%	4.46%
PR	461,782.68	USD 38,707.26	358,671.68	4.09%	3.63%	4.01%
SC	381,316.40	USD 31,897.36	295,614.30	3.37%	2.99%	3.30%
PE	296,525.39	USD 26,674.40	245,578.24	2.62%	2.50%	2.74%
CE	283,256.85	USD 26,306.65	241,516.32	2.51%	2.47%	2.70%
GO	211,737.70	USD 17,710.04	164,201.48	1.87%	1.66%	1.84%
DF	215,824.27	USD 17,586.84	163,481.93	1.91%	1.65%	1.83%
ES	178,018.10	USD 16,288.95	149,758.47	1.58%	1.53%	1.67%
PA	149,577.48	USD 15,217.73	138,630.39	1.32%	1.43%	1.55%
RN	127,776.91	USD 10,723.46	99,453.84	1.13%	1.01%	1.11%
MA	92,808.69	USD 9,469.19	86,230.52	0.82%	0.89%	0.96%
PB	85,136.65	USD 8,605.71	78,427.67	0.75%	0.81%	0.88%
AL	108,374.38	USD 8,542.69	79,756.64	0.96%	0.80%	0.89%
AM	84,106.32	USD 8,477.24	77,289.26	0.74%	0.80%	0.86%
MT	79,293.73	USD 6,947.26	64,070.08	0.70%	0.65%	0.72%
MS	78,910.04	USD 5,962.65	55,930.81	0.70%	0.56%	0.63%
PI	60,675.10	USD 5,641.81	51,775.17	0.54%	0.53%	0.58%
SE	52,311.76	USD 5,048.03	46,171.42	0.46%	0.47%	0.52%
RO	38,227.89	USD 3,418.13	31,482.69	0.34%	0.32%	0.35%
TO	26,700.81	USD 2,429.53	22,339.29	0.24%	0.23%	0.25%
AP	17,085.54	USD 1,656.10	15,153.00	0.15%	0.16%	0.17%
AC	12,466.44	USD 1,232.69	11,240.07	0.11%	0.12%	0.13%
RR	11,287.27	USD 1,211.90	10,988.08	0.10%	0.11%	0.12%
Total	11,299,547.97	USD 1,066,272.02	8,946,623.13	100.00%	100.00%	100.00%

In our example, we have a telco (Telefonica de Spana which operates predominately in São Paulo (SP), a state in Brazil. As we can see, Telefonica would be particularly well placed to offer a good deal for this particular organization, given the fact that we have 46.71 percent of the calls within the state of São Paulo. The graphic gives a better view:

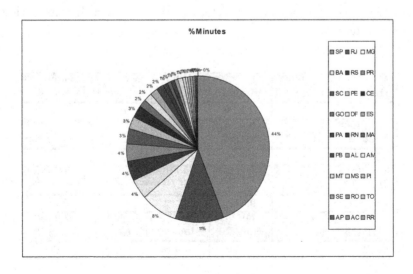

12.3   Traffic Hourly Distribution and Hourly Concentration Factor

We must determine the typical hourly distribution, which allows us to identify the hourly concentration factor: the percentage of daily traffic that occurs during the hour of heaviest traffic. Of course, doing this also identifies which hour has the highest volume.

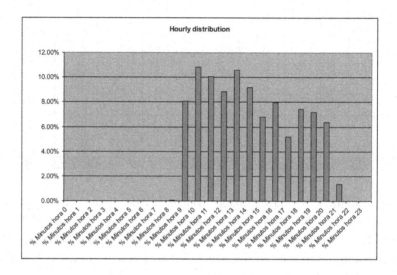

In this particular example, the HCF identified was 10.86 percent and occurred between nine and ten o'clock in the morning; the table below gives a better view:

% Minutes Hour 0	% Minutes Hour 1	% Minutes Hour 2	% Minutes Hour 3	% Minutes Hour 4	% Minutes Hour 5
0.00%	0.00%	0.00%	0.02%	0.00%	0.00%
% Minutes Hour 6	**% Minutes Hour 7**	**% Minutes Hour 8**	**% Minutes Hour 9**	**% Minutes Hour 10**	**% Minutes Hour 11**
0.00%	0.00%	0.03%	8.07%	10.86%	10.05%
% Minutes Hour 12	**% Minutes Hour 13**	**% Minutes Hour 14**	**% Minutes Hour 15**	**% Minutes Hour 16**	**% Minutes Hour 17**
8.86%	10.58%	9.17%	6.82%	7.98%	5.20%
% Minutes Hour 18	**% Minutes Hour 19**	**% Minutes Hour 20**	**% Minutes Hour 21**	**% Minutes Hour 22**	**% Minutes Hour 23**
7.45%	7.21%	6.34%	1.34%	0.01%	0.00%

Note that in this particular case, traffic outside typical business hours is almost nonexistent. Therefore, in a negotiation, discounts offered for calls outside business hours have no value whatsoever.

The calculation of this factor is crucial to enable us to do capacity planning. Based on this factor, we can calculate the bandwidth or the number of trunks necessary to support the traffic during the busiest hour.

12.4 Traffic Daily Distribution

This allows us to know how the traffic is distributed along the days during the month. Note that we have three curves: quant of connections, duration of the connections, and value.

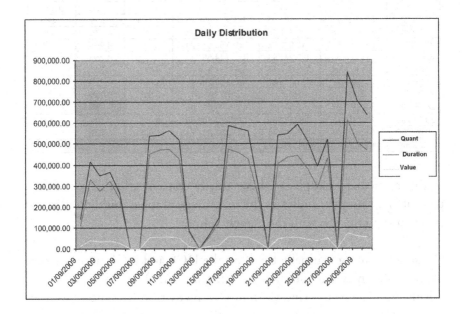

We may see the data through the spreadsheet:

Data	Quant	Minutes	Cost	% Quant	% Minutes MCF day	% Cost
01/09/2009	137,881.94	111,162.78	USD 13,208.01	1.22%	1.24%	1.24%
02/09/2009	416,640.32	331,844.86	USD 39,490.67	3.69%	3.71%	3.70%
03/09/2009	349,392.22	277,207.40	USD 34,265.18	3.09%	3.10%	3.21%
04/09/2009	366,275.72	320,608.93	USD 37,925.73	3.24%	3.58%	3.56%
05/09/2009	264,357.74	239,083.23	USD 27,829.30	2.34%	2.67%	2.61%
06/09/2009	30.10	88.63	USD 58.16	0.00%	0.00%	0.01%
07/09/2009	120.38	132.95	USD 95.27	0.00%	0.00%	0.01%
08/09/2009	538,436.29	451,327.60	USD 53,243.79	4.77%	5.04%	4.99%
09/09/2009	541,852.12	471,249.41	USD 55,099.04	4.80%	5.27%	5.17%
10/09/2009	564,453.74	474,538.90	USD 55,331.19	5.00%	5.30%	5.19%
11/09/2009	516,572.00	427,394.59	USD 50,016.69	4.57%	4.78%	4.69%
12/09/2009	85,034.47	78,955.39	USD 9,711.17	0.75%	0.88%	0.91%
13/09/2009	0.00	0.00	USD 0.00	0.00%	0.00%	0.00%
14/09/2009	64,088.09	52,107.67	USD 8,718.80	0.57%	0.58%	0.82%
15/09/2009	144,969.40	124,885.46	USD 16,441.05	1.28%	1.40%	1.54%
16/09/2009	589,417.85	474,336.32	USD 56,954.46	5.22%	5.30%	5.34%
17/09/2009	572,925.59	457,496.35	USD 56,664.31	5.07%	5.11%	5.31%
18/09/2009	560,240.39	428,711.40	USD 52,881.35	4.96%	4.79%	4.96%
19/09/2009	291,759.58	244,132.69	USD 29,838.14	2.58%	2.73%	2.80%
20/09/2009	0.00	0.00	USD 0.00	0.00%	0.00%	0.00%
21/09/2009	540,633.25	405,181.03	USD 47,831.93	4.78%	4.53%	4.49%
22/09/2009	547,856.14	435,467.65	USD 52,218.48	4.85%	4.87%	4.90%
23/09/2009	594,157.87	446,092.02	USD 53,570.24	5.26%	4.99%	5.02%
24/09/2009	511,019.41	380,241.41	USD 44,557.12	4.52%	4.25%	4.18%
25/09/2009	391,811.64	293,311.73	USD 34,942.02	3.47%	3.28%	3.28%
26/09/2009	521,703.27	431,376.67	USD 50,267.64	4.62%	4.82%	4.71%
27/09/2009	30.10	168.40	USD 121.46	0.00%	0.00%	0.01%
28/09/2009	842,218.97	612,941.92	USD 71,511.01	7.45%	6.85%	6.71%
29/09/2009	708,971.71	509,707.85	USD 59,588.09	6.27%	5.70%	5.59%
30/09/2009	636,697.68	466,869.75	USD 53,891.71	5.63%	5.22%	5.05%
Total	11,299,548.00	8,946,623.00	USD 1,066,272.00			

In this particular example, we have a daily concentration factor of 6.85 percent. This number indicates the percentage of the monthly traffic that occurs during the day with the biggest volume.

Capacity planning is usually based on the DCF and HCF (unless we are planning for new and non-already existent traffic); in other words, to calculate the necessary bandwidth or how many trunks we need, we have to consider our demand during the busiest hour of the busiest day of the busiest month (if there are different volumes along the months).

12.4 Call Duration Patterns

In regard specifically to voice traffic, if you have an active call center using dialers (automatic outbound call generators), we must constantly monitor the call duration pattern. This verification enables us to put in perspective the time of the contact and the associated cost. Calls with a duration of less than thirty seconds are called "short calls"; this usually occurs when the call is answered by a fax machine or answering machine and the dialer discharges the call without transferring it to the attendance group. Calls between thirty seconds and forty-five seconds may encompass those transferred to the attendance group but discharged before being attended (dropped in line). Monitoring these parameters allows you to know when there is a problem with the predictive dialer regarding call classification, dialing rate, or even the mailings themselves. Therefore, it is very important to have a clear view of the call duration distribution. The following spreadsheet is the kind of information telecom managers should have on a monthly basis.

From (Min)	to (Min)	Cost	Quant	Minutes	% cost	% quant	% minutes	% minutes accumulated
0.2	0.5	USD 333,259.53	7,528,854.90	3,127,686.46	31.25%	66.63%	34.96%	34.96%
0.5	0.7	USD 68,712.22	1,158,702.75	616,949.26	6.44%	10.25%	6.90%	41.86%
0.7	1	USD 72,477.99	804,376.07	590,209.32	6.80%	7.12%	6.60%	48.45%
1	1.5	USD 82,298.02	617,557.77	651,474.42	7.72%	5.47%	7.28%	55.73%
1.5	2	USD 72,589.58	388,355.13	574,566.59	6.81%	3.44%	6.42%	62.16%
2	3	USD 86,778.84	329,121.18	675,718.11	8.14%	2.91%	7.55%	69.71%
3	4	USD 56,144.43	148,550.65	431,692.09	5.27%	1.31%	4.83%	74.53%
4	5	USD 42,152.04	85,741.08	321,135.08	3.95%	0.76%	3.59%	78.12%
5	6	USD 33,251.08	54,782.92	251,174.44	3.12%	0.48%	2.81%	80.93%
6	7	USD 27,785.96	38,829.13	210,461.88	2.61%	0.34%	2.35%	83.28%
7	8	USD 24,344.35	29,592.73	185,230.25	2.28%	0.26%	2.07%	85.35%
8	9	USD 21,769.71	23,758.41	168,371.62	2.04%	0.21%	1.88%	87.24%
10	60	USD 144,223.73	91,261.48	1,138,089.82	13.53%	0.81%	12.72%	99.96%
60	100	USD 484.55	63.80	3,863.77	0.05%	0.00%	0.04%	100.00%
Total		USD 1,066,272.04	11,299,548.00	8,946,623.11				

The table is better understood through the following graphic:

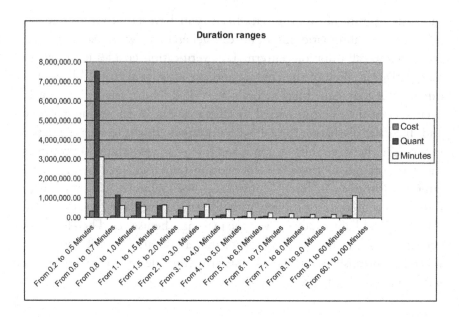

As we can see in this example, 76.86 percent of the calls have a duration equal to or less than 0.7 minutes, representing 37.70 percent of the cost. The average call duration is 0.79 minutes (47.50 seconds).

If we analyze the average cost of the contact, we may see a spreadsheet like the following:

Dutation range	Cost per call	Cost per minute
From 0.2 to 0.5 Minutes	USD 0.04	USD 0.11
From 0.6 to 0.7 Minutes	USD 0.06	USD 0.11
From 0.8 to 1.0 Minutes	USD 0.09	USD 0.12
From 1.1 to 1.5 Minutes	USD 0.13	USD 0.13
From 1.5 to 2.0 Minutes	USD 0.19	USD 0.13
From 2.1 to 3.0 Minutes	USD 0.26	USD 0.13
From 3.1 to 4.0 Minutes	USD 0.38	USD 0.13
From 4.1 to 5.0 Minutes	USD 0.49	USD 0.13
From 5.1 to 6.0 Minutes	USD 0.61	USD 0.13
From 6.1 to 7.0 Minutes	USD 0.72	USD 0.13
From 7.1 to 8.0 Minutes	USD 0.82	USD 0.13
From 8.1 to 9.0 Minutes	USD 0.92	USD 0.13
From 9.1 to 60 Minutes	USD 1.58	USD 0.13
From 60.1 to 100 Minutes	USD 7.59	USD 0.13

You should be able to get the real duration of the calls from your billing system, not only from the bill. This is necessary because if your charging granularity is for example thirty seconds, any call between zero and thirty seconds, regardless the real duration, will appear on the bill as lasting thirty seconds; the billing system, on the other hand, will indicate the real duration of the calls. We can group the billing logs by duration in ranges of six seconds and have a view like the following:

Range (secs)	Minutes	% minutes	Calls	% calls
00001 to 00006	46,389.81	3.43%	927,796.11	12.32%
00007 to 00012	551,633.78	40.10%	3,677,558.50	48.85%
00013 to 00018	312,629.99	18.40%	1,250,519.96	16.61%
00019 to 00024	344,884.91	20.14%	985,385.46	13.09%
00025 to 00030	309,417.69	17.93%	687,594.86	9.13%
Total (Bellow or equal 30secs)	1,564,956.17	100.00%	7,528,854.90	100.00%

Through this table, it becomes possible to verify that a typical short call lasts 12.47 seconds. This analysis makes it possible to determine which charging granularity suits the organization better:

Minutes charged 30+6+6	Minutes charged 6+6+6	Really used
3,127,686.46	2,258,656.47	1,564,956.17

In this particular case, through a granularity of 30s+6s+6s, the organization was charged a total of 3,127,686.46 minutes; we know that the average duration of the short calls is 12.47 seconds. Therefore, if we had a charging granularity of 6s+6s+6s, we would only pay for 2,258,656.47 minutes (of course, that would still be above the real usage of 1,564,956.17 minutes, but a lot better than using 30s+6s+6s granularity. In terms of value, this difference would be as follows:

3,127,686.46 × USD 0.12 = USD 333,259.52

2,258,656.47 × USD 0.12 = USD 271,038.78

That means a saving of USD 62,000.00 or 18 percent of the total spent on short calls, not a value to be despised.

12.4.1 Short Calls

The classical definition of a short call is a call whose duration wasn't long enough to involve any kind of treatment by the attendants. In an outbound call center, short calls include calls the dialer device identified as not valid, calls answered by a machine (answering machine or fax), or calls that were dropped by the client as soon as the salutation ended. In an inbound call center, short calls are those calls dropped before a certain time (usually after the salutation and soon after being placed in the attendance queue). Regardless of the precise definition, these calls don't generate a conversation.

We usually draw a line on this by time, arbitrarily defining short calls as those calls dropped by the outbound dialer before thirty seconds, no matter the reason (bad numbers, answering machines, etc.), and calls where a prospective customer hangs up before the conversation gets into full swing (between thirty seconds and one minute).

In an inbound call center, we may draw the line at calls dropped before the end of the salutation and calls dropped ten seconds after entering the attendance queue or ten seconds after being transferred to an agent (here we would have three types of short calls with different durations). It is very important to keep track of these calls, for several reasons:

1) It is important to know the proportion between the number of calls successfully concluded (total calls minus the short calls) and the number of calls tried (total calls). This information points to the efficiency of the process of reaching the client (or being reached by the client) in general, and in the case of outbound operations, the quality of the mailings used.

2) In a call center operation, it is important to know the cost of trying to reach the clients and the cost of actually talking with them. This information gives you an understanding of the cost of bad mailings.

3) It is important to know if the dialers are operating properly. A lack of balance between the number of tries and the number of successes can indicate that the predictive algorithm isn't working properly; the dialer may be producing more calls than the attendants can handle.

4) It is important to be able to compare the agents effectively worked time with the number of tries.

To really understand what is happening in your infrastructure, you are going have to dig deeper. Depending on the types of calls you are taking (or making), high volumes of short calls can be quite acceptable, especially in an outbound operation (wrong party contact, no interest, etc.).

If yours is primarily an inbound call center operation, it is worth understanding your call types and looking at the call stats at both a global and an individual level. Look for patterns or anomalies—high levels of transferred calls, individual agents with unusually high numbers of short calls, particular inbound numbers with high numbers of short calls—and drill down from there.

If you record the calls, identify a group of short calls and see if there are any similarities; if you have access to speech analytics, you can do this in volume, but even manually, you should get some insight. Some PBXs, and call recording programs, and call loggers allow you to see which party terminated the call. There are no industry averages; most call centers tend to have different experiences, even within the same industry. You need to think about what types of calls you take or make and consider what the likely customer expectation would be; this will give you some insight about what level of short calls are acceptable. It is very important to have monthly spreadsheets like one shown below:

time span (secs)	Calls	Minutes	Cost	% Calls	% accumulated calls	% minutes	% accumulated minutes	% cost	% accumulated cost
00000 a 00003	523,329.00	13,300.00	USD 70,545.91	2.15%	2.15%	0.06%	0.06%	1.82%	1.82%
00004 a 00006	782,036.00	66,857.00	USD 99,098.43	3.21%	5.37%	0.28%	0.34%	2.55%	4.36%
00006 a 00009	1,067,636.00	143,734.00	USD 90,747.57	4.39%	9.75%	0.61%	0.95%	2.33%	6.70%
00009 a 00012	3,741,951.00	728,751.00	USD 394,765.00	15.38%	25.13%	3.10%	4.06%	10.16%	16.86%
00013 a 00018	3,060,121.00	784,356.00	USD 207,004.50	12.58%	37.71%	3.34%	7.40%	5.33%	22.18%
00019 a 00024	2,314,978.00	827,882.00	USD 157,160.41	9.51%	47.23%	3.52%	10.92%	4.04%	26.23%
00025 a 00030	1,951,914.00	830,365.00	USD 128,368.77	8.02%	55.25%	3.79%	14.71%	3.21%	29.43%
00031 a 00060	6,358,839.00	4,513,260.00	USD 532,935.13	26.14%	81.39%	19.22%	33.93%	13.71%	43.21%
00061 a 00090	1,732,081.00	2,111,047.00	USD 279,953.31	7.12%	88.50%	8.99%	42.91%	7.20%	50.41%
00091 a 00120	861,690.00	1,498,263.00	USD 229,960.38	3.54%	92.05%	6.38%	49.29%	5.92%	56.33%
00121 a 00180	743,403.00	1,796,789.00	USD 327,999.88	3.06%	95.10%	7.65%	56.94%	8.44%	64.77%
00181 a 00240	325,213.00	1,124,499.00	USD 207,412.09	1.34%	96.44%	4.79%	61.73%	5.34%	70.10%
00241 a 00300	192,522.00	860,813.00	USD 149,395.70	0.79%	97.23%	3.66%	65.39%	3.84%	73.95%
00301 a 00600	387,239.00	2,690,770.00	USD 382,243.41	1.59%	98.82%	11.46%	76.85%	9.83%	83.78%
00601 a 00900	134,547.00	1,640,094.00	USD 219,260.94	0.55%	99.37%	6.98%	83.83%	5.64%	89.42%
00901 a 01200	67,353.00	1,161,435.00	USD 163,113.14	0.28%	99.65%	4.94%	88.78%	4.20%	93.62%
01201 a 09999	84,416.00	2,526,904.00	USD 242,861.56	0.35%	100.00%	10.76%	99.54%	6.25%	99.87%
10000 a 86400	512	108,849.00	USD 5,156.60	0.00%	100.00%	0.46%	100.00%	0.13%	100.00%
TOTAL	24,329,780	23,487,969.00	USD 3,886,602.73	100.00%		100.00%		100.00%	

12.5 Concentration of Traffic

Another key performance indicator (KPI) to be monitored is the traffic concentration per destination or origin. We must be able to see to or from how many numbers the calls were made, identifying how many numbers were called or called only one time, how many more than one time, and the distribution by quantity of calls (quantity grouped).

Outbound						
Item		Month				
	Quant		Minutes		Cost	
Calls made	19,703,828					
Numbers dialed	11,144,732	%	17,522,963.00	%	USD 2,491,412.03	%
Numbers dialed more than one time	4,049,393	36.33%	10,380,538.33	59.24%	USD 1,358,615.68	54.53%
Numbers dialed only one time	7,095,338	63.67%	7,142,424.67	40.76%	USD 1,132,796.35	45.47%
Cost associated with the repetition	12,608,490	63.99%	6,331,145.13	59.24%	USD 922,277.20	37.02%
Average number of calls per number (including the ones with one call)	1.77					
Average number of calls per number (only ones with repetition)	3.11					

This particular spreadsheet shows that 63.67 percent of the numbers called were called only one time. The average number of calls per number was 1.17, but if we remove the numbers called only one time, we have an average calls per number of 3.11.

As we can also see, 37.02 percent of the total cost is due to repetitive calls. The numbers called many times are usually numbers that belong to people related to the employees of the organization, suppliers, or clients. It is important to follow this KPI, because through it, you can spot several types of problems, from inadequate usage to technical problems. For example, if a private voice connection between headquarters and a specific site isn't operational, people may use the public network to reach this site (or the PBX may reroute the calls automatically); in this case, a high volume of calls to this specific number may raise a red flag.

Number of calls	Number of numbers dialled	Duration (minutes)	Cost	% quant	% Minutes	% Cost	Accumulated % Quant	Accumulated % Duração	Accumulated % Valor
1	7,095,338	7,142,424.32	USD 1,132,796.34	36.01%	40.76%	45.47%	36.01%	40.76%	45.47%
2	4,425,169	3,939,849.64	USD 531,962.30	22.46%	22.48%	21.35%	58.47%	63.24%	66.82%
3	2,484,202	2,054,360.13	USD 265,355.27	12.61%	11.72%	10.65%	71.08%	74.97%	77.47%
4	1,711,437	1,381,461.85	USD 177,508.65	8.69%	7.88%	7.12%	79.76%	82.85%	84.60%
5	1,071,103	850,256.18	USD 108,290.64	5.44%	4.85%	4.35%	85.20%	87.70%	88.94%
6	813,544	621,544.91	USD 79,877.35	4.13%	3.55%	3.21%	89.33%	91.25%	92.15%
7	528,488	398,229.04	USD 51,190.21	2.68%	2.27%	2.05%	92.01%	93.52%	94.20%
8	433,347	322,682.11	USD 41,202.40	2.20%	1.84%	1.65%	94.21%	95.37%	95.86%
9	272,597	199,272.19	USD 26,161.84	1.38%	1.14%	1.05%	95.59%	96.50%	96.91%
10	236,956	165,936.42	USD 20,873.14	1.20%	0.95%	0.84%	96.79%	97.45%	97.74%
11	157,522	110,147.52	USD 13,915.22	0.80%	0.63%	0.56%	97.59%	98.08%	98.30%
12	137,171	94,476.76	USD 11,744.94	0.70%	0.54%	0.47%	98.29%	98.62%	98.77%
13	89,111	63,799.71	USD 7,980.62	0.45%	0.36%	0.32%	98.74%	98.98%	99.09%
14	68,547	48,061.38	USD 5,968.05	0.35%	0.27%	0.24%	99.09%	99.26%	99.33%
15	44,211	30,844.25	USD 3,843.49	0.22%	0.18%	0.15%	99.31%	99.43%	99.49%
16	34,593	22,862.03	USD 2,848.25	0.18%	0.13%	0.11%	99.49%	99.56%	99.60%
17	23,240	16,420.99	USD 1,974.98	0.12%	0.09%	0.08%	99.61%	99.66%	99.68%
18	16,230	10,688.58	USD 1,322.33	0.08%	0.06%	0.05%	99.69%	99.72%	99.74%
19	12,526	8,793.09	USD 1,116.29	0.06%	0.05%	0.04%	99.75%	99.77%	99.78%
20	10,083	6,110.98	USD 758.36	0.05%	0.03%	0.03%	99.81%	99.80%	99.81%
21	7,126	4,944.92	USD 609.23	0.04%	0.03%	0.02%	99.84%	99.83%	99.83%
22	5,759	3,600.40	USD 442.77	0.03%	0.02%	0.02%	99.87%	99.85%	99.85%
23	3,345	2,696.84	USD 330.62	0.02%	0.02%	0.01%	99.89%	99.87%	99.87%
24	3,490	2,685.57	USD 328.17	0.02%	0.02%	0.01%	99.91%	99.88%	99.88%
25	4,121	2,590.28	USD 314.36	0.02%	0.01%	0.01%	99.93%	99.90%	99.89%
26	1,008	475.61	USD 58.76	0.01%	0.00%	0.00%	99.93%	99.90%	99.89%
27	1,571	800.47	USD 97.27	0.01%	0.00%	0.00%	99.94%	99.90%	99.90%
28	814	465.21	USD 61.98	0.00%	0.00%	0.00%	99.94%	99.91%	99.90%
29	562	467.81	USD 57.23	0.00%	0.00%	0.00%	99.95%	99.91%	99.90%
30	291	129.95	USD 24.09	0.00%	0.00%	0.00%	99.95%	99.91%	99.90%
31	601	382.91	USD 45.41	0.00%	0.00%	0.00%	99.95%	99.91%	99.91%
32	310	149.87	USD 16.26	0.00%	0.00%	0.00%	99.95%	99.91%	99.91%
33	640	315.34	USD 39.28	0.00%	0.00%	0.00%	99.96%	99.91%	99.91%
34	659	323.14	USD 57.23	0.00%	0.00%	0.00%	99.96%	99.92%	99.91%
36	349	160.27	USD 17.18	0.00%	0.00%	0.00%	99.96%	99.92%	99.91%
40	776	388.11	USD 65.66	0.00%	0.00%	0.00%	99.96%	99.92%	99.91%
48	465	409.77	USD 49.40	0.00%	0.00%	0.00%	99.97%	99.92%	99.92%
66	640	575.23	USD 74.56	0.00%	0.00%	0.00%	99.97%	99.92%	99.92%
70	679	307.54	USD 305.15	0.00%	0.00%	0.01%	99.97%	99.93%	99.93%
95	921	418.43	USD 44.65	0.00%	0.00%	0.00%	99.98%	99.93%	99.93%
110	1,066	506.79	USD 85.46	0.01%	0.00%	0.00%	99.98%	99.93%	99.94%
332	3,219	11,945.61	USD 1,596.64	0.02%	0.07%	0.06%	100.00%	100.00%	100.00%
Total	**19,703,828**	**17,522,962.13**	**USD 2,491,412.00**						

Note that in the previous spreadsheet, we can see how many numbers got how many calls; 7,095,338 numbers out of a total of 19,703,828 got only one call during this period. This view can help telecom managers understand the effectiveness of an active call center operation. In a corporate voice network, spot traffic concentration can identify potential savings. For example, if we identify a supplier whose traffic with the organization is really high, it may make sense to install a dedicated connection between the organization and the supplier.

In this organization, the analysis of traffic density showed that the 516,000 calls made during the month were designated to 241,700 numbers. When digging deeper, we identified the following:

Repetition	Numbers receiving the calls	Quant of calls	Minutes	Value
More than 50 calls month	574	52.792	106.342,60	USD 23.328,14
Between 20 and 50 calls month	1.908	58.440	112.879,80	USD 26.356,87
Bellow 20 calls month	239.230	404.855	672.383,60	USD 146.187,61
Total	241.712	516.087	891.606,00	USD 195.872,62

Repetition	Numbers receiving the calls	% Quant of calls	% Minutes	% Value
More than 50 calls month	0,24%	10,23%	11,93%	11,91%
Between 20 and 50 calls month	0,79%	11,32%	12,66%	13,46%
Bellow 20 calls month	98,97%	78,45%	75,41%	74,63%
Total	100,00%	100,00%	100,00%	100,00%

The spreadsheet shows that 574 numbers (0.24 percent) represented 11.91 percent of the whole cost, and 1.908 numbers (0.79 percent) represented 13.46 percent.

This fact is a strong indication that gains are attainable through analyzing why such a small quantity of numbers received such a large percentile of the calls made by the organization.

12.6 The Type of Traffic (Fixed-to-Fixed/ Fixed-to-Mobile)

In many countries, the tariff system differentiates between the costs associated with calls made among the fixed-line network (fixed-to-fixed), calls made among mobile lines (mobile-to-mobile), and calls made between these two networks (fixed-to-mobile or mobile-to-fixed).

This fact requires telecom managers in these countries to identify the volume and costs associated with each type of traffic. It is important to know the volume and cost associated with calls made to fixed lines and the volume and cost associated with calls made to mobile phones. This information should be identifiable, as shown below:

Type of service	Minutes	%	Cost	%
Fix-to-fix	8,480,408.01	94.79%	USD 663,442.60	62.22%
Fix-to-mobile	466,215.10	5.21%	USD 402,829.44	37.78%
Total	8,946,623.11	100.00%	USD 1,066,272.04	100.00%

In this particular example, calls destined to mobile phones represent 5.21 percent of the total but correspond to 37.78 percent of the costs. Usually calls between fixed lines and mobile lines (or vice versa) cost a lot more than calls between lines of the same type. There are several strategies that can be deployed to circumvent this tariff system peculiarity (GSM gateways, dedicated mobile trunks, etc.). However, identifying the traffic pattern is the first step.

12.7 Mapping the Mobile Traffic per Service Provider (Outbound)

When negotiating with a mobile provider, it is easier to get better prices for calls between lines that belong to that provider. These calls are usually called "on-net," meaning they don't leave the provider's network. Providers can usually be much more flexible in regard to the price of these calls, because they don't have to pay an interconnection fee (basically, what one provider has to pay another to transport a call originated in its own network and terminated in another).

In addition to that, the quality of calls transported over the same network (on-net) tends to be significantly better than those flowing through different networks.

These are the reasons why it is so important to map the mobile traffic by terminated call ("mobile traffic" meaning the traffic terminated in a mobile line) and inform the provider of the volume that could be on-net if it was originated by the provider's own trunks.

You may or may not use dedicated mobile trunks or GSM gateways. If you do, the need for mapping the traffic in regard to the terminated provider becomes even bigger, considering the fact that you may separate the outgoing mobile traffic by terminated provider, routing it through different trunk groups in order to make the call be between two trunks of the same provider. Proceeding in such a way guarantees that all mobile calls originate and terminate over the same provider (all mobile traffic becomes on-net). You basically identify the provider to which the call is destined and route the outbound call to a trunk group of this particular provider.

The deployment of GSM gateways may generate quality issues, depending on the density of the calls (quantity of calls generated per minute). This happens because in large outbound call centers deploying dialers, the number of calls generated exceeds the capacity of the base station controller (BSC) of the area. To avoid this kind of problem, you can disperse the calls through a network of dialers (distributed in different locations and using different BSCs) or install a direct link with the mobile provider.

Mapping traffic by termination point allows you to calculate the return on investment (ROI) of investing in routing devices and GSM gateways (needed to separate the outgoing traffic per mobile service provider). This strategy can be very effective in terms of reducing mobile traffic costs (it is not unusual to see reductions above 50 percent in some countries). The following spreadsheet shows an example of how this information should be presented:

Provider	Quant	Minutes	Cost	% Quant	% Minutes	% Cost
Telco 1	10,117.68	7,617.77	USD 5,047.10	1.62%	1.63%	1.25%
Telco 2	2,131.90	1,556.31	USD 1,491.75	0.34%	0.33%	0.37%
Telco 3	13,804.33	8,171.50	USD 7,826.88	2.21%	1.75%	1.94%
Telco 4	118,102.37	91,801.23	USD 78,499.55	18.90%	19.69%	19.49%
Telco 5	7,663.84	3,758.75	USD 3,595.70	1.23%	0.81%	0.89%
Telco 6	65,586.26	53,734.85	USD 47,914.91	10.50%	11.53%	11.89%
Telco 7	294.46	156.73	USD 149.97	0.05%	0.03%	0.04%
Telco 8	8,315.58	7,321.87	USD 7,000.83	1.33%	1.57%	1.74%
Telco 9	76,736.51	70,313.36	USD 58,461.07	12.28%	15.08%	14.51%
Telco 10	47.11	23.47	USD 13.34	0.01%	0.01%	0.00%
Telco 11	321,928.20	221,759.26	USD 192,828.34	51.53%	47.57%	47.87%
Total	**624,728.23**	**466,215.10**	**USD 402,829.44**			

The following spreadsheet shows the same information but separated by type:

VC1: Local
VC2: Intrastate
VC3: Interstate

Type	Quant	Minutes	Cost	% Quant	% Minutes	% cost
VC1	150,540.19	104,409.44	USD 56,757.51	24.10%	22.40%	14.09%
VC2	34,620.75	25,382.67	USD 24,265.06	5.54%	5.44%	6.02%
VC3	439,567.29	336,422.99	USD 321,806.85	70.36%	72.16%	79.89%
Total	**624,728.23**	**466,215.10**	**USD 402,829.42**			

In this particular example, it is clear that the relevant tariff is the interstate (VC3). It is also clear that provider Telco 11 terminates most of the calls. This information is crucial in any negotiation.

12.8 Routing Traffic

It is absolutely crucial to be able to route your traffic (this usually refers to outbound traffic but also applies to inbound). This provides three fundamental features:

- It allows you to forward calls to the provider that is able to transport them for the cheapest price.

- It allows you to forward calls through your own private networking, reducing the intraorganization costs and allowing you to use the public network at the nearest node (reducing long distance costs).
- It allows you to redirect calls in case of a major outage in a particular service provider.

Two key points when negotiating with the service providers are the ability to redirect calls to the provider able to transport it for the smallest cost and understanding traffic volume and interest (to where and from where the calls go).

If you know your traffic profile and interest but you are not able to route the traffic, your negotiation should try to get a lower tariff for the services with bigger volumes, at the same time avoiding high prices for the remaining services.

On the other hand, if you know your traffic and are able to route it, your negotiation can focus on getting the cheapest prices for specific services and let other services remain expensive (then you use your ability to redirect calls for these areas or services to go through other provider). Of course, in this case, you should carefully manage the minimum committed volumes.

When negotiating with service providers, the ability to route calls opens the possibility of extracting better tariffs for specific types of calls. This is crucial, given the fact that some regional service providers rely on others to deliver calls outside their own network. That means they usually charge less for the calls within their own network. Being able to route calls allows you to take advantage that and attain a much better overall telco cost.

Of course, being able to route traffic also makes it possible to deploy private networks with POPs. Without being able to forward the calls to specific trunks or tie-lines (least cost routing), it is not possible to establish POPs (from where the calls are collected—inbound traffic—or to where calls are forward to get the public network nearest to the destination—outbound traffic).

Another very important aspect associated with being able to route calls is the possibility of redirecting them in case of a major outage. This can be very important, especially if your operation is critical. It is particularly important when you have an outbound call center operation using dialers. Dialers demand some signaling adjustments, so it is hard to shift trunks quickly during an outage. Therefore, an effective backup requires you to have the trunks connected to the equipment and operating (with the signaling previously adjusted).

Once again, having at least two providers is highly recommended. Having only one provider, even if the provider has two different physical access infrastructures, will still expose the organization to administrative and negotiation problems that may affect the provider, such as strikes, bankruptcy, contracts disagreements, and so forth.

Example of Least Cost Routing

In this example, we are going to calculate how much an organization could be expending if it had the resources to route its traffic properly (these calculations do not take into account any discount). The analysis here goes like this: If you take your current contracts and route your traffic using your current prices, selecting the best alternatives in terms of cost, what would be your final telco cost? This particular organization had three contracts with three different service providers:

Telco 1
Telco 2
Telco 3

To execute the calculations properly, it was necessary to identify the typical standard of duration of the calls:

Type of call	%
Success calls	65.04%
Short calls	34.96%

Assuming short calls as those calls whose duration is equal to or below thirty seconds and taking as a reference only the fixed-to-fixed calls:

	Type	Telco 1	Telco 2	Telco 3
Success	Local	USD 0.040	USD 0.027	USD 0.044
calls	Long Distance within POP	USD 0.040	USD 0.090	USD 0.083
	Longa distance out of POP	USD 0.100	USD 0.090	USD 0.083
	Type	Telco 1	Telco 2	Telco 3
Short calls	Local	USD 0.020	USD 0.014	USD 0.022
	Long Distance within POP	USD 0.020	USD 0.027	USD 0.042
	Longa distance out of POP	USD 0.050	USD 0.027	USD 0.042

The item "Long Distance POP" corresponds to calls destined to locations where service provider Telco 1 has POPs and the calls are charged as local. Considering the traffic analyzed, the volumes are as follows:

		Average duration (secs)	Type	% tráffic	Minutes
	Success		Local	18.50%	1,019,769.06
	Calls	92	Long distance within POP	34.00%	1,874,170.17
Fix-to-fix			Long distance out of POP	47.50%	2,618,325.97
		Average duration (secs)	Type	% tráffic	Calls
8.480.408 Min.	Short Calls		Local	18.50%	2,196,426
		15	Long distance within POP	34.00%	4,036,674
			Long distance out of POP	47.50%	5,639,471

Considering the current tariffs, the costs would be as follows:

	Type	Telco 1	Telco 2	Telco 3
Success	Local	USD 40,790.76	USD 27,533.76	USD 44,517.00
Calls	Long distance within POP	USD 74,966.81	USD 168,506.64	USD 155,914.09
	Long distance out of POP	USD 261,832.60	USD 235,413.69	USD 217,821.16
	Type	Telco 1	Telco 2	Telco 3
Short Calls	Local	USD 43,928.51	USD 30,749.96	USD 47,941.38
	Long distance within POP	USD 80,733.48	USD 108,881.21	USD 167,907.48
	Long distance out of POP	USD 281,973.57	USD 152,113.46	USD 234,576.63
	Type	Telco 1	Telco 2	Telco 3
Combined	Local	USD 84,719.28	USD 58,283.72	USD 92,458.38
	Long distance within POP	USD 155,700.29	USD 277,387.85	USD 323,821.57
	Long distance out of POP	USD 543,806.16	USD 387,527.15	USD 452,397.79
	Total	**USD 784,225.73**	**USD 723,198.73**	**USD 868,677.74**

If we route the calls using the cheapest alternatives, we would have the following cost for fixed-to-fixed calls:

Type	Provider	Cost
Local	Telco 2	USD 58,283.72
Long distance within POP	Telco 1	USD 155,700.29
Long distance out of POP	Telco 2	USD 387,527.15
Total		**USD 601,511.16**

Here we can see that just routing the fixed calls properly can reduce the minimum operational cost from USD 723,198.73 to USD 601,511.16, a difference of USD 121,687.57 (17 percent). Now we are going to do the same calculations for fixed-to-mobile calls:

	Type	Telco 1	Telco 2	Telco 3	Telco 4
Success	VC1	USD 0.550	USD 0.650	USD 0.988	USD 0.395
Calls	VC2 e VC3	USD 0.550	USD 0.730	USD 0.988	USD 0.695
	Type	Telco 1	Telco 2	Telco 3	Telco 4
Short Calls	VC1	USD 0.270	USD 0.325	USD 0.494	USD 0.198
	VC2 e VC3	USD 0.270	USD 0.210	USD 0.494	USD 0.348

Using the current volumes:

		Minutes	Average duration (secs)	Type	% tráffic	Minutes
Fix-to-mobile	Success			VC1	22.40%	67,880.92
	Calls	303,039.82	92	VC2 and VC3	77.60%	235,158.90
		Calls	Average duration (secs)	Type	% tráffic	Calls
	Short Calls			VC1	22.40%	73,102.53
		326,350.57	30	VC2 and VC3	77.60%	253,248.04

We would have the costs as follows:

	Type	Telco 1	Telco 2	Telco 3	Telco 4	Telco 5	Gateway GSM
Success	VC1	USD 37,334.51	USD 44,122.60	USD 67,082.32	USD 26,812.96	USD 19,006.66	USD 12,897.37
Calls	VC2 e VC3	USD 129,337.39	USD 171,665.99	USD 232,392.32	USD 163,435.43	USD 89,360.38	USD 49,383.37
	Type	Telco 1	Telco 2	Telco 3	Telco 4	Telco 5	Gateway GSM
Short Calls	VC1	USD 19,737.68	USD 23,758.32	USD 36,121.25	USD 14,437.75	USD 10,234.35	USD 6,944.74
	VC2 e VC3	USD 68,376.97	USD 53,182.09	USD 125,134.33	USD 88,003.69	USD 48,117.13	USD 26,591.04
	Type	Telco 1	Telco 2	Telco 3	Telco 4	Telco 5	Gateway GSM
Combined	VC1	USD 57,072.188	USD 67,880.919	USD 103,203.568	USD 41,250.712	USD 29,241.011	USD 19,842.115
	VC2 e VC3	USD 197,714.364	USD 224,848.083	USD 357,526.648	USD 251,439.128	USD 137,477.509	USD 75,974.413

If we execute the calls using the cheapest alternatives, we would use Telco 5 or GSM gateways to transform fixed-to-mobile calls into mobile-to-mobile.

In this way, we can calculate the minimum possible cost if we route the calls through the cheapest alternative. In this particular example, just doing that could save 27.95 over the current expenditures, without any tariff negotiation or implementation of GSM gateways.

Type	Least cost routing	Current cost	Savings	% savings
Fix-to-fix	USD 601,511.16	USD 663,442.00	USD 61,930.84	9.33%
Fix-to-mobile	USD 166,718.52	USD 402,829.00	USD 236,110.48	58.61%
Total	USD 768,229.68	USD 1,066,271.00	USD 298,041.32	27.95%

These kinds of calculations are easily executable, and you should be able to do them regularly.

Very often, the difficulties associated with properly routing traffic are linked to the unavailability of trunks or routing devices; nevertheless, you should always be aware of your potential gains if you have routing resources.

12.9 Planning for Expansion

We need a very clear process to define the planning of the network; the telecommunications area should be included in discussions about new applications and defined processes to identify changes in demand and determine how to address those changes.

12.10 Capacity Planning

The telecommunications area should be attentive to changes in demand, trying to anticipate those changes. Through monitoring tools, the telecommunications area should be able to analyze whether the resources available are adequate. Over-/undercapacity situations need to be identified before user services are adversely affected.

The telecommunications area should also participate in discussions about new services and applications, indicating to the organization the technical and economic impact of adding or removing services. It is also crucial to have processes in place that give the telecommunications area some sort of insight about site expansions and organic growth. This should be done within a defined process, not sporadically (telecom committee meetings are a good example of how this can be done).

Physical expansion of the network is often an interesting situation, as it is common for organizations to expand by acquisition. These are often not announced internally to an organization until the deal is done. In these cases, the telecommunications area needs to be part of the IT due diligence function, as a large acquisition could have a significant impact on both usage and cost base of the telecommunications infrastructure.

12.11 Evaluating New Technologies

This is not a daily process, but there can be a consistent approach to evaluating new technologies. The activities that are part of this evaluation must be documented and consistently executed to keep current in the technological world. It is a matter of due diligence of the telecommunications area to scan the market, looking for products and solutions that may add value to the organization. Technical staff should have access to magazines and be able to attend conventions where they can compare their own experience with other professionals. It is very important that the organization provide resources for this function. Otherwise, the influence of service providers and hardware vendors will grow to a point that is not in the organization's best interest.

The telecommunications area is also responsible for training the users of the telecommunications resources, letting them know how to make the most of the available resources.

Chapter 13: Mobile Device and Mobile Applications Management

Management of mobile devices and management of mobile services are usually addressed separately. However, the way we see the issue, these are just two sides of the same coin and therefore should be treated jointly. Typically, the term "mobile device" encompasses the following:

- mobile phones
- smartphones
- tablets
- mobile POS
- mobile printers

That said, we have to keep in mind that managing the inventory of mobile devices and services in a large corporation is an important part of managing the whole inventory of telecom resources. However, managing mobile devices and services has some additional challenges:

- Mobile devices, as the name makes clear, are not statically located in a defined place and therefore can't be easily associated with an address. Mobile devices have to be associated with the user responsible for it.
- Mobile devices support several services, with several charging strategies and prices.
- Mobile devices have several user profiles, including different sets of services that each user has access to.
- Mobile devices have several limits for usage per service per user.

All these particularities make managing mobile devices and services a much harder task than managing other resources.

The following diagram shows the basic entities to be controlled when dealing with mobile devices and services and how they are interrelated:

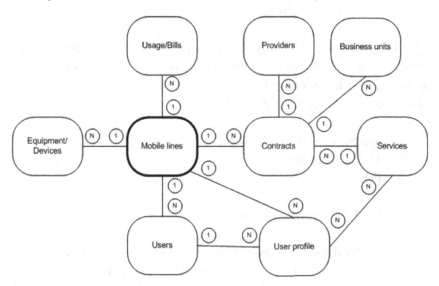

As we can see, mobile lines are associated with other entities with the following logic:

- A mobile device may be associated with several mobile lines (some devices support one, two, or more mobile lines), but a mobile line is associated with only one device.
- A user may be associated with more than one mobile line, but a mobile line is associated with only one user.
- A bill may be associated with more than one mobile line, but a mobile line is associated with only one bill.
- A user profile may have several users associated with it, but a user has only one profile.
- A contract may have several mobile lines associated with it, but each line is associated with only one contract.
- A service (price list) may be associated with only one contract, but a contract may have several services, with several charging strategies and prices.

- A contract may be associated with only one provider, but a provider may have several contracts.
- A contract may be associated with only one business unit, but a business unit may have several contracts.
- A contract may support several services with several charging strategies (price list).

This diagram makes it clear that managing mobile devices and services encompasses many more variables than managing other devices. In addition, mobile devices include at least three items that don't need to be controlled in other types of devices:

- user name and identification
- SIM card number
- IMEI

1) **User name:** Each mobile device is associated with a specific user.
2) **SIM card number:** A subscriber identity (or identification) module (SIM) is an integrated circuit that securely stores the international mobile subscriber identity and the related key used to identify and authenticate subscribers on mobile telephony devices (such as mobile phones and computers).
3) **IMEI:** The International Mobile Station Equipment Identity (IMEI) is a (usually unique) number that identifies 3GPP (i.e., GSM, UMTS, and LTE) and iDEN mobile phones, as well as some satellite phones. It is usually found printed inside the battery compartment of the phone, but it can also be displayed on-screen on most phones by entering *#06# on the dialpad, or viewed along with other system information in the settings menu on a smartphone.

The main control point is to track which user is responsible for each device and mobile line, which services are associated with the line and with which profile (which usage limits), and of course to which contract and tariffs the line is associated.

Besides those three exclusive attributes, it is also advisable to keep track of the following aspects of mobile devices:

- operating system, including version
- software installed
- restrictions blocking specific software
- restrictions to software functionalities
- access control to the devices (passwords or special access codes)
- data communication through Wi-Fi
- data communication thorough Bluetooth
- camera activation
- anti-virus program

Controlling these items enhances the security of the devices, reduces the risk of unavailability, and keeps down technical support costs. It is worth noticing that outdated versions of some operating systems can open the door for "treats" of all sorts, including viruses and hacker attacks. Therefore, maintaining a tight control over device configurations will improve security and reliability.

The operational model adopted by the organization can include providing a mobile device or not; users may be allowed to use their own devices: BYOD (Bring Your Own Device).

In order to make the control process easier, we should group the services into packets of services or user profiles, associating them to hierarchical levels within the organization. For example:

Operational User (Profile 1)

- voice services, 200 minutes per month
- no web access
- no data packet
- no SMS packet
- free intragroup calls

Managers (Profile 2)

- voice services, 500 minutes per month
- web access but limited to 1000 megabits per month
- 100 SMS messages per month
- free intragroup calls

Directors (Profile 3)

- unlimited number of minutes per month
- unlimited web access
- unlimited SMS messages
- free intragroup calls

These profiles (which are specific to each organization) reflect the usage policy adopted by the organization. The recommended strategy here is to frame every line into one of the predefined profiles. The main point is to standardize profiles to make the management of services and lines easier.

Although the mobile resource managing tools made available by providers usually allow service parameters to be set in a line-by-line basis, in a large organization, it isn't advisable to proceed in this way. Managing a large number of lines with a plethora of associated services without a standard is very time intensive and complicated.

13.1 Defining the Usage Policies and Usage Packet/ Profiles

It is crucial to properly define usage profiles. The profiles must be structured in a way where that the specific needs of each type of user are properly covered. A well-defined policy should reduce costs and at the same time enhance user performance.

A good policy must encompass the following aspects:

- eligibility

- plans of service and usage limitations by profile
- strategy of device deployment
- reimbursement policies
- contingency plans for loss or theft of the device
- maintenance policies

13.1.1   Eligibility

Mobile devices and services are working tools; using them properly can enhance the performance of the organization as a whole and reduce its operational costs.

But to achieve these objectives (enhance performance and reduce costs), we have to properly define the needs of each type of person (employee or not) who is going to use them.

Therefore, before anything else, we need to define who is eligible to have a mobile line. Subsequently, we need to define what service pack would support the needs of each type of user (this definition also includes the type of device provided).

The evaluation of user needs must include who needs what, including voice, SMS, chat, Internet, specific tools, and applications.

A thorough analysis should be conducted based on the historical usage of each hierarchical level of the organization. As mentioned before, the idea is to establish some sort of usage pattern per hierarchical level (if possible).

The result of this analysis is a set of profiles indicating who is eligible to have a mobile resource and with which associated services.

13.1.2   Plans of Service and Usage Limitations

Before negotiating a mobile services contract, it is crucial to already have the users' profile defined, indicating who needs what and how many users will be in each profile. This information enables you to determine the volume needed, which is the basis for negotiation with providers.

Usually, there is a tradeoff between minimum usage and the cost per unit used. Therefore, knowing the volume (including the self-imposed limitations in usage per user profile) is the key to negotiating good tariffs.

The profiles defined must adhere to the configuration possibilities offered by the provider's management tool. For example, there is no point defining a user profile whose maximum monthly usage is 100 minutes if there is no possibility of configuring the service with such a limitation.

On the other hand, you should try as much as possible to reconcile your own internal user profile with the predefined plans offered by the providers. This means that even if a provider has seemingly lower prices per service unit, you may achieve a lower cost overall by choosing a provider with a charging structure more adherent to your needs.

A very common mistake is beginning the process from without instead of within, looking first at the service plans offered by the providers and then at the organization's real needs.

13.1.3   Strategy of Device Deployment

Ownership of the device is one of the more disputed issues when defining a mobile services policies.

From the user perspective, the root cause of the issue is the fact that today, people may be able to directly get mobile devices as good as, or even better than, the ones provided by the organization (not to mention enhanced services plans). In summary, it is not uncommon to find users with better devices than the ones provided by the organization, using a service plan they see as more adherent to their needs.

From the organization's perspective, there are two forces at play: 1) The need for management and standardization of the devices, with all its obvious benefits, which pulls to the adoption of a policy where the devices are provided by the organization, and 2) the cost to maintain the devices (if users bring their own, this would be the users' responsibility).

On top of these two perspectives, there is a widespread perception that users are better at choosing the devices and service plans that adhere to their needs, although most are not as cost conscious as they would be if the costs were not reimbursed by the organization. In other words, if the users choose the devices and the plans, their productivity tends to increase, but so do the costs—and not necessarily in the same proportion.

As a general rule, the more valuable the user time is, the more flexible the organization can be in regard to these two issues (device and plan). This happens because even a small increase in the productivity of a high value employee may be worth more than the additional mobile bill.

13.1.4   Reimbursement Policies

An important part of the mobile policy is to define how the cost will be paid. Of course, if all devices belong to the organization and all service plans are corporate plans, all bills will be paid directly by the organization (even in this scenario, we may have situations where the user has to reimburse the organization for some nonauthorized services; for example, private long distance calls).

But if users contract the plan directly, you may have situations where the organization reimburses the users (totally or partially).

Another common scenario is when we have a corporate plan with zero cost for communication among the plan's users. In that scenario, the organization sometimes provides lines not only for the employees themselves but also for their relatives. The rationale for this strategy is that most non-work-related calls are to relatives, and this saves the organization money on those calls. On the other hand, the organization may only pay the bills of the lines used directly by the employee. The other lines have to be paid directly by the relatives of the users. The benefit for the users is that they may get a better tariff than they would be able to get directly (and also save the cost of calling their relatives). This arrangement is more complex and demands a more complete set of policies and better management.

13.1.5 Contingency Plans for Loss or Theft of the Devices and Maintenance Policies

If the organization provides the mobile devices (not BYOD), it is imperative to have a clear policy defining how the losses of the devices will be treated. Here it is important to draw a distinction between the types of loss, trying to enhance the responsibility of the users in regard to the device entrusted to them. Therefore, policies should include clear procedures, and it is imperative to define parameters within which the devices will be replaced (entirely or partially) by the organization. For example, users who lose more than one device a year may be responsible for the cost of the second device lost.

The same logic should be applied to maintenance; we should clearly define the criteria within which devices are replaced due to maintenance problems, defining how many times the organization will be responsible for fixing the device without charging the user. It is very important to make users know they have to be careful with the device entrusted to them.

13.1.6 Management Tools

The definition of policies only makes sense if you can configure and enforce them. This is the reason why mobile device management (MDM) tools are so important. These tools allow you to secure, monitor, and manage mobile devices deployed across different mobile operators and service providers.

MDM functionality typically includes over-the-air distribution of applications, data, and configuration settings for all types of mobile devices, including mobile phones, smartphones, tablet computers, ruggedized mobile computers, mobile printers, and mobile POS devices. This applies to both company-owned and employee-owned (BYOD) devices across the organization.

To effectively deploy a mobile device management tool, the telecom management team should understand the specifications of the tool and its implementation.

13.1.7   Device Management Specifications

Mobile device management programs enable corporate IT departments to control the many mobile devices used across the enterprise; therefore, when specifying a MDM tool, some key points should be considered:

Ability to implement the service profiles: The management tool should allow the configuration of the service profiles defined by the usage policies of the organization. This configuration should apply to a single mobile device, an entire fleet of mobile devices, or any IT-defined set of mobile devices.

Independent platform: The management tool should meet the common definitions of an open standard, meaning the specification is freely available and implementable. It should be supported by several mobile devices, such as PDAs and mobile phones. The Open Mobile Alliance (OMA) specifies a platform-independent device management protocol called OMA Device Management.

Remote configuration: Over-the-air (OTA) programming capabilities are considered a main component of mobile network operator and enterprise-grade mobile device management software. These capabilities include the ability to

- remotely configure a single mobile device, an entire fleet of mobile devices, or any IT-defined set of mobile devices;
- remotely send software and OS updates;
- remotely lock and wipe a device, which protects the data stored on the device when it is lost or stolen; and
- troubleshoot remotely.

13.1.8 Implementation

The proper implementation of a management tool is the cornerstone of an effective mobile device control program. Having an inventory of devices and services indicating the service profile of each device and user is basic in the whole process. The tool has to allow the implementation of these profiles, but you have to have the list of resources and users, including the profiles, to be able to configure them.

Once again, there is no sense in defining usage profiles if you can't configure them into the devices. Therefore, the profiles must adhere to the tool's configuration possibilities. Typical solutions include a server component, which sends out the management commands to the mobile devices, and a client component, which runs on the handset and receives and implements the management commands. In some cases, a single vendor may provide both the client and the server; in others, clients and servers will come from different sources.

The management of mobile devices has evolved over time. At first, it was necessary to either connect to the handset or install a SIM in order to make changes and updates; scalability was a problem. Today, most tools allow a client-initiated update, similar to when a user requests a Windows update.

Most mobile management tools use commands sent over the air. An administrator can use an administrative console to update or configure any one handset, group, or groups of handsets. These scalability benefits are particularly useful when the number of managed devices is large.

Through mobile management tools, device management centers ensure that users benefit from plug-and-play data services for whatever device they are using. Such a platform automatically detects devices in the network and sends them settings for immediate and continued usability. The process is fully automated, keeps history of used devices, and only sends settings to subscriber devices that were not previously set, sometimes at speeds reaching fifty over-the-air setting updates per second.

Chapter 14: Risk Management

Telecommunications usually takes a back seat when an organization evaluates the risks of its IT infrastructure; many organizations simple don't realize how dependent they are on telecommunications. Therefore, an adequate assessment of the risks associated with losing the telecom infrastructure (totally or partially) is crucial. The organization has to understand thoroughly the impact the lack of connectivity (voice and data) can have in each aspect of its operations.

Once we assess the risk associated with losing connectivity, we have to define a backup strategy. Backup strategies are very important, because they allow us to define the reaction capability of the structure. When designing a telecom network, we must define what is supposed to happen when each one of its components fail. Of course, the level of reliability of a structure depends on the organization's activity, and the business must be able to assess the costs associated with the lack of connectivity (downtime).

We cannot access risk in the telecommunication network in an isolated way, focusing exclusively on the telecom aspect. Therefore, if we assess the telecom risk and discover that the organization is critically dependent on having connectivity, we should determine whether the best strategy is to increase the reliability of the network or reduce the dependence of the organization on it.

Consequently, if an organization that only operates on-line implements resources to allow it to operate off-line (even if this off-line operation isn't exactly the same as when it is on-line), this organization will reduce its exposure to downtimes in the telecom network. This strategy may be better and cost less than simply increasing the network's reliability.

To know which strategy is best, we need to know how much the downtime costs. Based on the downtime cost, we can identify the maximum costs a backup can incur in order to avoid make the backup cost more than the loss of connectivity.

Here we have to identify from which value we consider it worth implementing a backup structure. This is important because even after identifying the maximum cost of the backup, it wouldn't make sense to implement a backup structure whose cost would be exactly the same of the loss due to lack of connectivity. It has to cost less than this threshold. The difference between the loss due to lack of connectivity and the actual cost of the backup is the gain due to the existence of the backup.

Once we identify this value, it becomes the basis over which several backup alternatives will be considered. In other words, we must be able to identify which backup alternatives are available within the identified budget. However, we should be aware that very often, identifying how much the downtime cost isn't an easy task. Nevertheless, an honest effort must be done to identify it, preceding any backup planning.

14.1.1   Points to Consider when Increasing the Reliability of the Telecom Structure

Define who provides the main and backup networks: The providers of the main circuit/trunks and the backup ones must be different. Organizations that don't do that are exposed to two sets of problems:

- A technical problem affecting the main circuit may affect the backup.
- Even if the backup infrastructure is really a separated technical structure, this arrangement doesn't protect the organization against administrative problems with the provider's operations such as bankruptcy and strikes.

We should be attentive to the fact that the last mile of the main circuit/ trunks and the backup must be different and belong to different providers;

this is an important point, because even if you contract with two separate providers, they may share the physical access to the site.

Define how many alternatives there are for each resource: Depending on how critical the connectivity of the site is, you may have more than one alternative in case of resource failure.

Define how the switching process will occur: There must be a mechanism to switch from one provider to another smoothly and reliably, without adding critical failure points.

Define the technology of the main and backup networks: In critical sites, it is prudent to deploy different sets of technologies for the main and backup links (for example, fiber and satellite).

Define if the backups will be kept empty: You may decide to only activate the backups in case of failure of the main alternative, or they may be used on a regular basis (with spare capacity). In general, "on use" backup structures tend to be more reliable than "just in need" ones. Resources that are not normally used have a much bigger chance to be defective when you need them. Furthermore, if you are using resources that otherwise would be empty, you are also saving money and keeping them monitored.

The downside of the "on use" backup is that there is a tendency to lose track of the real volumes being transported by the backup links. As time passes, you may have a situation where the importance and volume of the traffic transported by the backup links grows. In extreme cases, the failure of the backup can become a risk in itself, or the volume transported by the two links cannot be compressed into one of them, and at the same time, all applications end up having a similar level of criticality.

The ideal scenario is having the flows divided evenly between the main and the backup links (50/50), and in case of one failure, the traffic is 100 percent transported by the remaining one. This 50/50 proportion doesn't include applications that can be stopped in case of need, such as voice or downloading of files.

Define the scenario of a major failure of a provider: Create a scenario where we remove all connections belonging to the service providers chosen in the main network; doing that forces us to look the structure as if the whole main provider went down (macro level).

Define the functionality of the devices: You also should be careful in regard to concentrating functionalities in a few devices, which tends to make the structure cheaper and simplify the operation; however, concentrating the whole connectivity of a site over a few devices can generate critical points of failure.

Therefore, we must define how the functionalities will be distributed per type of device, when and how new technical environments will be implemented, and which strategy of backup will be deployed. If we don't define these points in advance, what usually happens is that such definitions end up being made by default, considering only the circumstances of the moment.

Define the technical environments: Another important aspect that directly affects the risk of the structure is the issue of multiple technical environments and the lack of standardization of these environments. The general rule guiding our decisions should be to minimize the number of environments and standardize these environments, not only in terms of brands but above all in terms of functionalities per type of resource. Although a bit obvious, always remember that standardization tends to reduce training, maintenance, management, and support costs; it also increases productivity.

The separation in many technical environments (whether in the same site or not) should be done in such a way as to guarantee some level of interchangeability between them. It should be done in order to guarantee that in case of problems with one environment, there is the possibility of handling at least part of the traffic through the other. This is what we call a "hot structural backup" for the environment as a whole.

Define the maintenance policy: It is important that the organization have a *defined* policy in regard to maintenance. In general, it is highly advisable to have permanent maintenance contracts in large installations.

The absence of a maintenance contract exposes the organization to downtimes; even if there is the possibility of paying the hardware provider on a per-visit basis, that may imply a lack of priority in a moment of need. Usually, large and critical operations have permanent maintenance contracts, and these contracts include very strong SLAs and preventive check-ups. As mentioned, the need for permanent maintenance contracts grows as the size, criticality, and complexity of the infrastructure grows. When contracting maintenance, some aspects should be observed:

- The basic reference to define the monthly value to be charged is the value of the equipment. In this context, the price paid monthly corresponds to a kind of insurance. The normal value usually spins between 0.5 and 1.5 percent.
- The maintenance provider must have all necessary hardware for ready substitution in case of need (passive reserve). Eventually, a cheaper contract may be negotiated, transferring the responsibility of keeping backup hardware to the organization (it is cheaper but not advisable); very large organizations may use this modality.
- Any contract must include periodic configuration backup. The backup should be stored off-site by the service provider.
- Precise SLAs should be defined, indicating the maximum response time, including diagnosis and fixing of the problem as well as maximum resolution time after maintenance request. The contract should also indicate exactly how this time is going to be counted. Usually, most organizations deploy only one backup for each link.

Examples of backup strategies:

- The main sites and all aggregation nodes will have at least two data circuits, one primary and one secondary (backup), belonging to different providers and using separate last miles.
- The main sites and all aggregation nodes will be connected to at least two central offices (COs) of each service provider, and arrangements are made to guarantee that in case of problems in one of the COs, the traffic will flow entirely though the other.
- The main sites and all aggregation nodes will be provided with batteries to support the telecommunication structure for two

hours and electrical generators with enough fuel to run for twelve hours.

Case: Evaluating the Cost Benefit of a Backup Structure in a Retail Chain

This case discusses the process of assessing the effectiveness of the backup structure in a retail organization encompassing 125 stores. The cost of the backup links is USD 76,825 month. Adding to the economic aspect of the backup analyzed in this case, the organization also had three technical issues that limited the effectiveness of the backup structure:

- The backup links are connected to the same equipment as the main link; that leaves the CPE as a single point of failure within the store, despite the existence of two links in each store.
- The backup gateway (in the organization's data center) doesn't support the whole volume of the traffic in case the telco providing the main links goes down (a rare but possible event).
- Today seventeen stores (13 percent of the total) operate without backup links.

The point that triggered the analysis was the fact that those seventeen stores were operating without backups for more than two years. Therefore, the understanding we sought was fourfold:

- What is the correlation between the economic loss associated with lack of connectivity per hour versus the cost of the backup versus frequency and duration of the failure events?
- What could be done to reduce the loss?
- What could be done to reduce the cost of the backup?
- What could be done to reduce the frequency and duration of the downtime?

In this case, we will focus on the first two bullets. Our main objective is to establish the estimated loss due to the absence of a backup and compare it with the cost of having a backup. Subsequently, we will try to identify strategies to reduce the economic losses associated with lack of connectivity.

In this particular organization, the identification of these parameters was made easier by the fact that we had a sample to compare with (the seventeen stores).

The operation of the stores was dependent on connectivity for sales made with credit cards or bank cards; those amounted to almost 80 percent of all sales. However, that doesn't mean that when the store was without connectivity, all credit and bank card sales were lost. The sales system was designed to allow sales using credit cards and bank cards in off-line mode (for purchases amounting to less than USD 50.00). This value encompasses almost 70 percent of all sales using bank or credit cards.

These sales were done without verification, and there was a loss of approximately 5 percent of the value sold in these circumstances. Therefore, in a off-line operation, something around $0.8 \times 0.7 \times 0.95 = 0.53 + 0.20 = 0.73 = 73$ percent of all sales would be executed and be valid. That leaves us with a revenue reduction of 27 percent when the store is off-line.

Another important factor comes to play: the average profitability of the organization; in our specific case, 8 percent. That means the organization obtained USD 8.00 per each USD 100.00 in sales.

Considering the average working hours of a store of ten hours and knowing the amount sold by each store per month, we can make an average sales per hour (and per POS), and the loss of sales per hour and the loss of real revenue per hour of off-line operation.

With this information, we evaluate the probability of the store going off-line, which is achieved based on the history of real events. To do this calculation, we sum the time down during the working hours of the stores during a month divided by the sum of the working hours of the stores. That brings us to a scenario where 99.8 percent of the time, the stores are on-line. A very low probability; nevertheless, we calculated the cost of each hour down in each store, if this event actually happened.

That brings us to the sales volumes of the stores:

Size of the store	Monthly revenue	Daily revenue	Hourly revenue On-line)	Hourly revenue off-line (73%)	Raw loss for being off-line (1 hour)	Real revenu loss for being off-line (1 hour)
Small	USD 200.000,00	USD 8.000,00	USD 800,00	USD 584,00	USD 216,00	USD 17,28
medium	USD 500.000,00	USD 20.000,00	USD 2.000,00	USD 1.460,00	USD 540,00	USD 43,20
Large	USD 1.000.000,00	USD 40.000,00	USD 4.000,00	USD 2.920,00	USD 1.080,00	USD 86,40

Considering the fact that each backup link costs around USD 700, we would need to have an outage as follows to justify the cost of the backup:

Real revenu loss for being off-line (1 hour)	Backup link costing USD 700 - Number of hours down to justify the cost	Considering the total working hours month of 200. % of down time	Up-time
USD 17,28	40,51	20,25%	79,75%
USD 43,20	16,20	8,10%	91,90%
USD 86,40	8,10	4,05%	95,95%

This spreadsheet makes it clear that we would need to have an up-time of just 79.75 percent to justify the implementation of a backup with this cost in a small store; considering our real scenario, where the historical up-time is 99.8 percent, it doesn't seem to make sense. The same goes for the other types of stores, although the discrepancies diminish as the size of the store grows.

Of course, here we made an oversimplification, considering the sales evenly distributed along the month and the day, but even if we consider aspects such monthly and daily concentration of traffic, the value doesn't grow enough to change the conclusion. The conclusion is simple: The backup in its current value and strategy of use isn't worth its cost.

Faced with this fact, the organization decided to deploy backup links only in the large stores, but even this decision was made after considering another factor. The organization decided to deploy the backup links for other services (voice, video training, and downloading of files); in that case, the cost of having the backups would now be shared between the loss for lack of connectivity and the benefits of these other services. These services can be interrupted if the need for backup arises.

In this particular example, the calculation of the loss associated with the lack of connectivity is quite straightforward; sometimes, it isn't. Nevertheless, we must strive to get as much rationality as possible in this process.

Another point this case demonstrates is the importance of having some sort of strategy to deal with the lack of connectivity. If we had a situation where all sales using credit cards and bank cards were held in case of lack of connectivity, we would have a scenario where only 20 percent of the sales would be made when the store is off-line. In that case, we would have a loss per hour down as follows:

Size of the store	Monthly revenue	Daily revenue	Hourly revenue On-line)	Hourly revenue off-line (20%)	Raw loss for being off-line (1 hour)	Real revenu loss for being off-line (1 hour)
Small	USD 200.000,00	USD 8.000,00	USD 800,00	USD 160,00	USD 640,00	USD 51,20
medium	USD 500.000,00	USD 20.000,00	USD 2.000,00	USD 400,00	USD 1.600,00	USD 128,00
Large	USD 1.000.000,00	USD 40.000,00	USD 4.000,00	USD 800,00	USD 3.200,00	USD 256,00

That shows a scenario where just 2.73 hours down in a large store costs the organization more than the cost of the link (5.46 and 13.67 for medium and small stores, respectively). That brings our uptime spreadsheet to the following:

Real revenu loss for being off-line (1 hour)	Backup link costing USD 700 - Number of hours down to justify the cost	Considering the total working hours month of 200. % of down time	Up-time
USD 51,20	13,67	6,84%	93,16%
USD 128,00	5,47	2,73%	97,27%
USD 256,00	2,73	1,37%	98,63%

This particular case makes it clear that strategies that reduce dependence on connectivity can be more cost effective than increasing reliability.

Chapter 15: Reports and Analysis

In order to be able to manage the telecom infrastructure, we need to define a set of key performance indicators (KPIs) through which we follow the traffic and the cost. These KPIs are followed through reports. These reports are the means through which we see what is going on with the organization's traffic and telecom costs. In general, the reports encompass two types of information: financial information and traffic information.

The first step is to identify the destination and the source of the information. That is crucial, because by identifying what information is necessary and where it can be found, we can build an effective control process. There are typically four sources of information:

- the telecom resource management system (usually a dedicated software but it can also be a spreadsheet)
- the billing system
- the bills auditing system
- the data links monitoring tools

The destination of the information is usually the following:

- users (administration and call centers)
- accounts payable and accounting departments
- department of financial planning (budgeting)
- TI and telecom area

The challenge is to identify which information each area needs and how to provide it from the already mentioned sources.

15.1 Users (Administration and Call Center)

Users can usually be divided between administrative users and call center users. Each has a different set of needs in terms of information.

15.1.1 Administrative Users

The administrative users usually have their information needs linked to the identification of quantity, duration, and cost of calls per extension or user. Usually this kind of information is also consolidated by department and cost center. The objective of these reports is to inhibit the spurious use of the telephone system and register the internal costs properly.

The reports destined to the administrative users are usually generated from the billing system, and the area managers are usually the ones who receive and analyze it. Such reports usually are generated monthly in a department basis. Usually, department managers receive a report indicating the number of calls, number of minutes, and cost of each extension (or user) associated with their department (or cost center). Typically, the analysis of such reports is done by the manager directly. However, it is important to have in place guidelines and policies orientating the manager about what is acceptable and what isn't and how to proceed in case of no compliance (how to proceed in case of reimbursements, for example). Therefore, it is crucial to have a defined policy orientating the managers.

In some parts of the world, organizations implement what is called prepaid extensions (or users). That means defining the maximum amount of minutes that each extension (or user) can use per month. When the user reaches the maximum, the system blocks further use; any additional usage above the predefined value has to be approved by management (this case highlights the importance of telecom costs in some countries).

Thoroughly controlling the use of voice resources can potentially reduce the number of calls to the absolutely necessary; it can also reduce call duration. Usually this kind of effort has a potential to reduce the traffic by 10 to 15 percent.

15.1.2   Call Center Operations

The call center operations usually need more detailed reports than the ones required by administrative users. This is because the level of detailing about the traffic necessary to manage a call center is greater.

15.1.2.1 Tracking the Volume

Usually the basic report to manage the traffic/cost of a call center operation represents the quantity of calls with duration and cost. The calls would be separated by type and by trunk group. A typical report would be structured as follows:

Inbound traffic:

- traffic inbound from mobile phones, local
- traffic inbound from mobile phones, long distance
- traffic inbound from a fixed line, local
- traffic inbound from a fixed line, long distance

Outbound traffic

- traffic outbound to mobile phones, local
- traffic outbound to mobile phones, long distance
- traffic outbound to a fixed line, local
- traffic outbound to a fixed line, long distance

These reports usually show the quantity of calls, minutes, and cost per type of call per trunk group. Usually we have columns with the values of the last twelve months.

Jan/12						Feb/12					
Trunk group	Type	Quant	Minutes	Value charged	Value calculated (auditing)	Trunk group	Type	Quant	Minutes	Value charged	Value calculated (auditing)
32162000	Long distance - Fi	148.015	190.346,10	$ 22.051,11	$ 11.993,92	32162000	Long distance - Fi	90.033	116.334,10	$ 13.526,72	$ 7.830,39
32162000	Long distance - Mo	14.312	13.111,20	$ 9.782,12	$ 5.928,86	32162000	Long distance - Mo	12.003	11.798,70	$ 8.814,74	$ 5.333,49
32162000	Local - Fi	1.039	1.091,10	$ 464,97	$ 29,06	32162000	Local - Fi	2.492	2.644,40	$ 1.513,89	$ 70,21
32162000	Local - Mo	4.218	3.921,10	$ 1.898,43	$ 1.803,77	32162000	Local - Mo	4.222	4.014,30	$ 2.973,74	$ 1.646,53
32392000	Long distance - Fi	375.390	487.203,00	$ 36.579,98	$ 30.699,44	32392000	Long distance - Fi	380.617	468.124,30	$ 57.918,00	$ 31.387,91
32392000	Local - Mo	4.698	4.220,30	$ 3.128,33	$ 1.941,28	32392000	Local - Mo	7.842	7.472,90	$ 3.335,37	$ 3.437,44
32392000	Local - Fi	2.309	2.813,10	$ 1.395,11	$ 74,81	32392000	Local - Fi	1.997	1.927,20	$ 651,38	$ 51,47
32392000	Long distance - Mo	43.394	40.042,40	$ 28.808,05	$ 18.120,82	32392000	Long distance - Mo	44.723	42.438,90	$ 42.280,64	$ 20.006,28
35132500	Local - Fi	4.394	4.412,30	$ 327,44	$ 167,22	35132500	Local - Fi	40	39,70	$ 2,31	$ 1,69
35132500	Long distance fixo	73	112,90	$ 27,28	$ 11,82	35132500	Long distance fixo	223	338,10	$ 80,79	$ 34,99
35132500	Local - Mo	2.857	2.424,20	$ 2.247,99	$ 1.401,77	35132500	Local - Mo	19	20,20	$ 17,83	$ 11,12
35132500	Long distance - Mo	65.507	34.005,50	$ 30.531,82	$ 31.902,36	35132500	Long distance - Mo	177.000	149.042,80	$ 138.861,46	$ 86.457,90
35132500	Não data hado	0	190.657,78	$ 73.328,89	$ 73.328,89	35132500	Não data hado	142.534	173.010,80	$ 86.505,40	$ 86.505,40
Total		666.801	954.361,08	$ 254.791,56	$ 179.004,22	Total		863.247	1.007.181,40	$ 353.484,67	$ 242.474,51

15.1.2.2 Call Duration Reports

Besides the basic volume report, the proper management of a call center operation requires other types of reports. An important one is the report tracking the duration of the calls. This report shows the percentage of the calls within a duration range. It allows us to verify the percentage of the calls deviating from what would be the typical duration (productive call). This deviation may be up (long calls) or down (short calls). This important analysis enables us to see the correlation between number of calls and effective contacts and effective contacts and costs. In addition, this report allows us to spot problems caused by dialers, telco network signaling, and interactive voice response devices IVRs.

These reports should separate the traffic inbound and outbound (separation by trunk group is also advisable). These reports are based on the bill logs and also on the billing system logs. The following example shows how the calls are distributed through the duration ranges (this specific report was based on the outbound bill logs):

Range of duration	Quant	Minutes	Cost	% Quant
bellow 6 sec	0	0,00	R$ 0,00	0,00%
beween 6 and 12 sec	0	0,00	R$ 0,00	0,00%
beween 12 and 18 sec	0	0,00	R$ 0,00	0,00%
beween 18 and 24 sec	0	0,00	R$ 0,00	0,00%
beween 24 and 30 sec	0	0,00	R$ 0,00	0,00%
beween 30 and 36 sec	407.946	203.973,00	R$ 106.162,17	34,27%
beween 36 and 42 sec	23.429	14.057,40	R$ 7.451,16	1,97%
beween 42 and 48 sec	20.875	14.612,50	R$ 7.645,66	1,75%
beween 48 and 54 sec	20.172	16.137,60	R$ 8.554,37	1,69%
beween 54 and 60 sec	19.104	17.193,60	R$ 9.022,06	1,61%
beween 60 and 90 sec	424.941	453.009,60	R$ 76.109,27	35,70%
beween 90 and 120 sec	98.611	165.872,50	R$ 51.388,71	8,29%
beween 120 and 150 sec	52.007	113.090,90	R$ 37.110,04	4,37%
beween 150 and 180 sec	28.086	75.200,70	R$ 26.741,09	2,36%
beween 180 and 210 sec	18.284	58.154,10	R$ 22.484,13	1,54%
beween 210 and 240 sec	11.257	41.491,40	R$ 16.358,56	0,95%
beween 240 and 300 sec	15.795	69.651,40	R$ 29.236,40	1,33%
beween 300 and 360 sec	11.631	63.320,70	R$ 26.594,93	0,98%
beween 360 and 420 sec	10.531	67.792,60	R$ 28.977,69	0,88%
beween 420 and 480 sec	8.521	63.271,20	R$ 26.842,21	0,72%
beween 480 and 540 sec	6.061	51.044,00	R$ 21.559,70	0,51%
beween 540 and 600 sec	4.026	37.917,20	R$ 16.102,41	0,34%
beween 600 and 900 sec	6.902	80.650,50	R$ 34.412,82	0,58%
beween 900 and 1800 sec	1.899	35.541,90	R$ 14.828,43	0,16%
beween 1800 and 6000 sec	141	5.386,30	R$ 2.242,23	0,01%
Total	**1.190.219**	**1.647.369,10**	**R$ 569.824,05**	

The minimum charging granularity adopted by telcos is usually thirty seconds; any call whose duration is equal to or below thirty seconds is charged as lasting thirty seconds. Because of that, all calls whose duration is equal to or below thirty seconds are grouped in the line "between 30 and 36s." The range of six seconds is arbitrary.

If the same analysis was done based on the billing system logs, where the real duration of the calls were registered, the volume concentrated in the range between thirty and thirty-six seconds would be distributed through the inferior ranges.

In this particular example, we should note that having a charging granularity of 30s in a operation where 34 percent of the calls are short calls is highly unfavorable to the organization.

Another aspect is that 34 percent is a very high percentage of short calls. That points to the existence of problems in the devices connected to the public network in regard to call classification (probably a public network signaling identification). The difficulty in recognizing if a number is valid or not is the core of the problem. When the devices connected to the public network fail to identify that a number isn't valid, they fail to drop the call before four seconds, which is usually the minimum time required for a call to be considered valid and charged by the telco. That means the organization will pay for all tries, even when the number isn't valid. A similar problem happens when the call reaches an answering machine in a mobile line.

The following graphic shows the duration distribution of a typical organization:

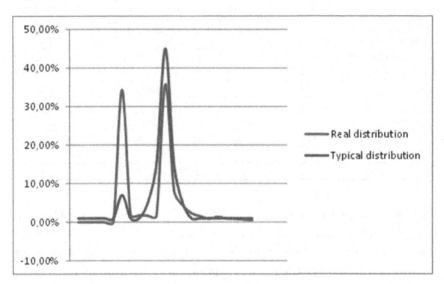

Through this graphic, we can see clearly that there is a problem with the percentage of short calls.

Tracking the inbound call duration distribution is also important, given the fact that through it, we can spot problems with the users' access to the call center or with attendants' quality.

Range of duration	Quant	Minutes	Cost	% Quant	% custo
bellow 6 sec	0	0,00	R$ 0,00	0,00%	0,00%
beween 6 and 12 sec	0	0,00	R$ 0,00	0,00%	0,00%
beween 12 and 18 sec	0	0,00	R$ 0,00	0,00%	0,00%
beween 18 and 24 sec	0	0,00	R$ 0,00	0,00%	0,00%
beween 24 and 30 sec	0	0,00	R$ 0,00	0,00%	0,00%
beween 30 and 36 sec	110.114	55.057,00	R$ 29.950,69	11,65%	5,48%
beween 36 and 42 sec	3.464	2.078,40	R$ 1.123,02	0,37%	0,21%
beween 42 and 48 sec	3.519	2.463,30	R$ 1.326,85	0,37%	0,24%
beween 48 and 54 sec	2.806	2.244,80	R$ 1.229,58	0,30%	0,22%
beween 54 and 60 sec	3.008	2.707,20	R$ 1.476,60	0,32%	0,27%
beween 60 and 90 sec	364.982	374.750,80	R$ 28.611,52	38,63%	5,23%
beween 90 and 120 sec	40.829	69.439,80	R$ 11.521,40	4,32%	2,11%
beween 120 and 150 sec	38.284	83.878,40	R$ 13.609,43	4,05%	2,49%
beween 150 and 180 sec	31.596	85.148,40	R$ 13.062,01	3,34%	2,39%
beween 180 and 210 sec	30.293	96.931,40	R$ 14.363,36	3,21%	2,63%
beween 210 and 240 sec	29.685	109.821,50	R$ 16.457,18	3,14%	3,01%
beween 240 and 300 sec	57.657	256.042,80	R$ 38.998,84	6,10%	7,13%
beween 300 and 360 sec	49.350	267.953,80	R$ 41.408,25	5,22%	7,57%
beween 360 and 420 sec	38.620	248.371,80	R$ 39.126,82	4,09%	7,15%
beween 420 and 480 sec	29.640	220.169,60	R$ 35.318,75	3,14%	6,46%
beween 480 and 540 sec	23.159	195.182,50	R$ 31.961,77	2,45%	5,84%
beween 540 and 600 sec	18.231	171.990,10	R$ 28.373,71	1,93%	5,19%
beween 600 and 900 sec	44.783	534.764,60	R$ 91.539,60	4,74%	16,74%
beween 900 and 1800 sec	22.067	431.487,10	R$ 83.518,10	2,34%	15,27%
beween 1800 and 6000 sec	2.773	105.909,80	R$ 23.957,75	0,29%	4,38%
Total	**944.860**	**3.316.393,10**	**R$ 546.935,24**		

Here we can see clearly the high price paid by this specific organization for calls with a duration above what would be typical (inbound traffic). That can be an indication of problems with the quality of the services or problems with the signaling with the public network (call hold).

The subsequent graphic shows a comparison between the distribution of the inbound call duration of a specific organization and a typical distribution:

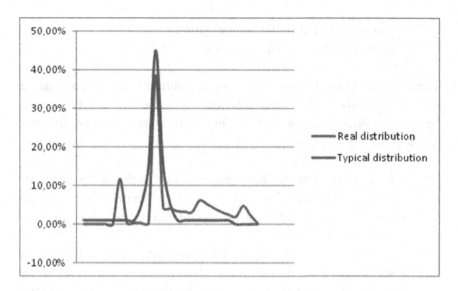

As we can see, 11.65 percent of the inbound calls have a duration equal to or smaller than thirty seconds, which may indicate a problem with the IVR menu. We also can see that 44.04 percent have a duration above 120 seconds.

In general, only 44 percent of the inbound calls were treated within the time we consider typical (costing only 6.38 percent of the total).

The situation described above is a clear indication of possible problems with receiving the calls and terminating them.

In this specific organization, these reports show that there is a high percentage of short calls in the outbound traffic and a high percentage of short and long calls in the inbound traffic.

Such calls represent a high percentage of the costs. Both issues can be associated with problems with the public network signaling and with the call classification by the devices.

15.1.2.3 Call Concentration Reports

There are two other important reports:

- concentration of traffic by destination
- concentration of traffic by source

These reports identify the number of calls, minutes, and cost associated with each number called (outbound traffic) and each number received (inbound traffic) during a month. They address several parameters that can help track the quality of service:

1) First call resolution: This indicator shows us that many users had their problem solved in the first call. Through this report, we map how many users had to call the call center one, two, three, or more times during a month.

2) Repetition: This indicator shows how many times the same user was accessed by the call center during a month. This indicator is important as a reference about the effectiveness of the dialing process and the quality of the mailing.

15.1.2.3.1   Inbound Traffic Concentration

The following spreadsheet shows traffic concentration by origin in an inbound operation:

inbound Call-center					
	December 2012				
Item	Quant				
Calls received	944.870		Value		
Numbers of origim	321.407	%	R$ 546.935,24	%	
Numbers of origin with more than one call	197.311	61,39%	R$ 436.853,65	79,87%	
Numbers of origm with only one call	124.096	38,61%	R$ 110.081,59	20,13%	
Repetition cost	623.463	65,98%	R$ 330.061,98	60,35%	
Average number of calls per number including all numbers	2,940				
Average number of calls per number for numbers with more than one call	4,160				

The previous spreadsheet shows a traffic profile where only 38.61 percent of the users access the call center only one time during the month and 61.39 percent do it more than one time. Note the high level of repetition (average of four calls per user who called more than one time).

The issue becomes more interesting when we distribute the calls and percentages in quantity ranges:

170

Quantity of calls per number	Quantity of numbers	Minutes	Value	% numbers	% Minutes	% Value
1	124.096	660.441,10	R$ 97.873,54	38,610%	19,907%	17,89%
2	72.634	562.626,70	R$ 82.204,17	22,599%	16,958%	15,03%
3	43.101	443.043,50	R$ 67.989,00	13,410%	13,354%	12,43%
4	26.158	339.271,50	R$ 54.080,97	8,139%	10,226%	9,89%
5	16.683	259.041,20	R$ 42.590,55	5,191%	7,808%	7,79%
6	11.004	201.747,50	R$ 34.055,04	3,424%	6,081%	6,23%
7	7.273	152.206,60	R$ 26.805,89	2,263%	4,588%	4,90%
8	5.242	124.452,00	R$ 22.785,89	1,631%	3,751%	4,17%
9	3.670	97.773,00	R$ 18.226,96	1,142%	2,947%	3,33%
10	2.560	74.399,60	R$ 14.356,36	0,796%	2,243%	2,62%
More than 10 calls	8.986	402.686,10	R$ 85.966,88	2,796%	12,138%	15,72%

As we can see, 2.8 percent of the numbers that called this particular call center did so more than ten times during the month. Such numbers account for 12.13 percent of the spoken minutes and 15.72 percent of the total cost.

If we apply the concept of first call resolution, we note that only 38.61 percent of the users call this particular call center only one time during the month, and these calls represent just 17.89 percent of the cost (we have a toll-free inbound call center).

15.1.2.3.2 Outbound Traffic

The subsequent spreadsheet shows the traffic concentration by destination in an outbound call center:

Outgoing Call-center						
		Dec/12				
Item		Quant			Value	
Calls made		1.190.276				
Numbers of origim		594.474	%		R$ 569.824,05	%
Numbers of origin with more than one call		241.405	40,61%		R$ 334.072,05	58,63%
Numbers of origim with only one call		353.069	59,39%		R$ 235.752,00	41,37%
Repetition cost		595.802	50,06%		R$ 183.248,53	32,16%
Average number of calls per number including all numbers		2,002				
Average number of calls per number for numbers with more than one call		3,468				

The previous spreadsheet shows that, in this particular call center, 59.39 percent of the numbers called are called only one time during the month. The average repetition for the numbers called more than one time is 3.46 times.

Quantity of calls per number	Quantity of numbers	Minutes	Value	% numbers	% Minutes	% Value
1	353.069	580.334,80	R$ 231.539,80	59,392%	35,224%	40,63%
2	116.638	337.234,60	R$ 126.084,62	19,620%	20,469%	22,13%
3	52.191	208.243,00	R$ 70.852,23	8,779%	12,639%	12,43%
4	27.081	136.908,40	R$ 42.494,47	4,555%	8,310%	7,46%
5	15.398	92.766,90	R$ 26.366,12	2,590%	5,631%	4,63%
6	9.482	66.669,50	R$ 17.424,38	1,595%	4,047%	3,06%
7	5.968	48.085,30	R$ 11.681,63	1,004%	2,919%	2,05%
8	4.062	36.632,50	R$ 8.885,12	0,683%	2,223%	1,56%
9	2.733	27.464,30	R$ 6.186,23	0,460%	1,667%	1,09%
10	1.975	21.830,50	R$ 5.119,55	0,332%	1,325%	0,90%
More than 10 calls	5.877	91.390	R$ 23.190	0,989%	5,547%	4,07%

As we can see in the previous spreadsheet, 24.38 percent of the cost is concentrated in the numbers with more than three calls.

15.2 Accounts Payable and Accounting Departments

These two areas are important demanders of information from the telecom area. This information should be provided monthly, with the invoices to be paid indicating the value charged and the value approved. The information should also include the values associated with each cost center and ledger account in each invoice. The association between the resource, the value, and the cost center and ledger account has to be made by the telecom area, because it is the only area within the organization where there is control over where each resource is installed and who actually uses it.

Therefore, the telecom area has to be able to forward the list of invoices to be paid (already approved), indicating the total value of the invoice and the values within each invoice associated with each cost center and ledger account.

Invoice Number	Service Provider	Contract Description	Payment Date	Due Date	Total Generated	Total Charged	Total Approved
17301	DESKTOP	ACESSO DEDICADO INTERNET - CD MONTE MOR	7/30/2013	7/30/2013	R$ 1,299.90	R$ 1,299.90	R$ 1,299.90
650243	TELEFONICA EMPRESAS	Single	7/12/2013	7/28/2013	R$ 99.00	R$ 99.00	R$ 99.00
15004263	EMBRATEL	PRIMELINK - ORBITAL	7/17/2013	7/25/2013	R$ 2,544.24	R$ 2,544.24	R$ 2,544.24
0999B	TELEFONICA EMPRESAS	LINKS DE DADOS - TELEFONICA	7/30/2013	7/25/2013	R$ 16,206.53	R$ 16,206.53	R$ 16,206.53
999	TELEFONICA EMPRESAS	LINKS DE DADOS - TELEFONICA	7/30/2013	7/25/2013	R$ 870,915.11	R$ 830,329.27	R$ 830,329.27
529536B	VIVO EMPRESAS	CONTRATO VIVO CELULARES	7/25/2013	7/25/2013	R$ 3,679.53	R$ 3,679.53	R$ 3,679.53
529536C	VIVO EMPRESAS	CONTRATO VIVO CELULARES	8/1/2013	7/25/2013	R$ 4,279.75	R$ 4,279.75	R$ 4,279.75
34721	VIVO EMPRESAS	CELULARES VIVO DDD 21	7/12/2013	7/25/2013	R$ 36.00	R$ 369.52	R$ 369.52
37214	VIVO EMPRESAS	CELULARES VIVO DDD 31	7/11/2013	7/25/2013	R$ 0.00	R$ 118.11	R$ 118.11
36968	VIVO EMPRESAS	CELULARES VIVO DDD 41	7/16/2013	7/25/2013	R$ 0.00	R$ 2,559.60	R$ 2,559.60
4621	VIVO EMPRESAS	CELULARES VIVO DDD 67	7/12/2013	7/25/2013	R$ 104.80	R$ 2,588.69	R$ 2,588.69
529536	VIVO EMPRESAS	CONTRATO VIVO CELULARES	7/25/2013	7/25/2013	R$ 18,448.10	R$ 68,344.58	R$ 68,344.58
529539	VIVO EMPRESAS	PLACAS 3G CORPORATIVAS	7/25/2013	7/25/2013	R$ 9,640.55	R$ 10,494.19	R$ 10,494.19
8939	EMBRATEL	Single	8/8/2013	7/24/2013	R$ 251.63	R$ 251.63	R$ 251.63
192	CTBC-ALGAR	CENTRAL DE RELACIONAMENTO- POP DF	7/29/2013	7/20/2013	R$ 0.00	R$ 35.07	R$ 35.07
15241	CTBC-ALGAR	CENTRAL DE RELACIONAMENTO- POP GO	7/29/2013	7/20/2013	R$ 0.00	R$ 148.56	R$ 148.56
234646	CTBC-ALGAR	CENTRAL DE RELACIONAMENTO- POP MG	7/29/2013	7/20/2013	R$ 0.00	R$ 1,237.68	R$ 1,237.68
2601	CTBC-ALGAR	CENTRAL DE RELACIONAMENTO- POP MS	7/29/2013	7/20/2013	R$ 0.00	R$ 38.13	R$ 38.13
276	CTBC-ALGAR	CENTRAL DE RELACIONAMENTO- POP PR	7/29/2013	7/20/2013	R$ 0.00	R$ 38.13	R$ 38.13
401	CTBC-ALGAR	CENTRAL DE RELACIONAMENTO- POP RJ	7/29/2013	7/20/2013	R$ 0.00	R$ 38.13	R$ 38.13
91475	CTBC-ALGAR	CENTRAL DE RELACIONAMENTO-NVM	8/1/2013	7/20/2013	R$ 0.00	R$ 61,570.13	R$ 61,570.13
161	OI BRT TNL	DATA LINK ORBITAL	7/30/2013	7/20/2013	R$ 2,613.04	R$ 2,753.69	R$ 2,753.69
106342	OI BRT TNL	VOZ REGIAO BRT MATO GROSSO DO SUL	7/30/2013	7/19/2013	R$ 112.22	R$ 112.22	R$ 112.22
91111	OI BRT TNL	VOZ REGIAO BRT MATO GROSSO DO SUL	7/30/2013	7/19/2013	R$ 1,525.56	R$ 1,525.56	R$ 1,525.56
59258	CTBC-ALGAR	VOZ CTBC LOJAS DE SAO PAULO	7/11/2013	7/15/2013	R$ 866.71	R$ 1,666.81	R$ 1,666.81
211000016	EMBRATEL	Single	7/5/2013	7/15/2013	R$ 723.24	R$ 723.24	R$ 723.24
Total					R$ 933,345.91	R$ 1,013,051.89	R$ 1,013,051.89

15.3   Department of Financial Planning (Budgeting)

The telecom area has to be able to provide the organization's financial planning area with reliable forecasts about the telecom costs (usually the next six to twelve months). These forecasts should be based on the historical values combined with the foreseen effects of the processes already being implemented by the organization (for example, expansions [cost up] and contract negotiations [cost down]). These reports are usually provided once or twice a year.

15.4   TI and Telecom Area

One of the main responsibilities of a telecom area in a large organization is to keep physical and financial control over the resources used. That is necessary not only to provide information to other areas but also to make possible the management of the resources, including moves and changes, payment approval, and identification of cost-reduction opportunities. To manage a telecom infrastructure properly, we have to have in place a set of key performance indicators to follow the traffic and the costs. These KPIs should be followed through a set of reports. They are the means through which we see what is going on with the organization's traffic and telecom costs.

15.4.1   Financial Reports

The management of telecom in a large organization has as its main goal guaranteed connectivity between the users, with the desired quality of service for the least cost possible. Therefore, cost management is a fundamental part of managing telecom in a large organization, and telecom managers have to be cost conscious. They have to understand that part of their job is to find ways to get more for less from the infrastructure under their responsibility. It is not only a matter of making things work, but making things work for less.

Due to this fact, the first step is to have a clear understanding of your organization's actual telecommunications expenditures, including telco expenditures, hardware, and services. It is also important to be able to discriminate the expenditures, identifying not only the services themselves but also things such as penalties for overdue payments, penalties for not achieving minimum committed traffic, and so forth.

Using financial reports, we can follow the costs by several criteria. A fundamental point in a financial report is to be able to see the historical values. It is the only way to put the information in perspective. This is an important consideration, because many people tend to emphasize the precision of the numbers when controlling expenses and bills. In our view, although there is a need for auditing bills and checking the values, the main point is to follow the historical values and determine if there is any significant deviation from the average value (up or down). We also have to be able to see the costs by several criteria:

- by service provider (what we refer to as a baseline report)
- by service provider by business unit
- by business unit
- by contract
- by resource
- by cost center

As mentioned before, all these reports should show the values of the last twelve months. The first step is to build a baseline report, indicating the total telecom expenditures during the last twelve months. The following spreadsheet is an example of this type of report:

Service provider name	Jan-12	Feb-12	Mar-12	Apr-12	May-12	Jun-12	Jul-12	Aug-12	Sep-12	Oct-12	Nov-12	Dec-12
CLARO EMPRESAS	CAD 22,118.87	CAD 2,983.31	CAD 1,879.20	CAD 1,440.06	CAD 662.72	CAD 0.00	CAD 0.00	CAD 0.00	CAD 0.00	CAD 0.00	CAD 0.00	CAD 0.00
CTBC-ALGAR	CAD 62,938.40	CAD 106,726.84	CAD 163,628.42	CAD 122,929.68	CAD 136,700.42	CAD 165,590.50	CAD 161,776.75	CAD 147,563.27	CAD 28,425.92	CAD 67,568.23	CAD 55,442.40	CAD 4,488.96
EMBRATEL	CAD 56,133.41	CAD 54,036.88	CAD 62,463.19	CAD 65,820.70	CAD 52,736.90	CAD 81,895.52	CAD 60,762.26	CAD 147,650.70	CAD 173,855.01	CAD 213,532.97	CAD 226,948.74	CAD 247,384.26
GIGACOM DO BRASIL LTDA.	CAD 10,179.79	CAD 10,179.79	CAD 10,179.79	CAD 10,179.79	CAD 10,179.79	CAD 10,179.79	CAD 10,179.79	CAD 10,179.79	CAD 10,179.79	CAD 10,179.79	CAD 10,179.79	CAD 10,179.79
NEXTEL	CAD 1,355.75	CAD 1,269.80	CAD 1,391.47	CAD 1,251.15	CAD 1,243.10	CAD 1,015.60	CAD 851.03	CAD 1,053.26	CAD 940.17	CAD 941.75	CAD 940.13	CAD 938.94
OI BRT TNL	CAD 68,067.45	CAD 66,319.42	CAD 63,820.44	CAD 47,961.75	CAD 91,672.92	CAD 69,297.86	CAD 66,179.46	CAD 60,010.15	CAD 53,835.36	CAD 50,574.59	CAD 52,062.57	CAD 51,983.75
SERCOMTEL	CAD 2,123.88	CAD 1,863.27	CAD 1,843.35	CAD 1,751.86	CAD 1,866.33	CAD 2,022.70	CAD 1,574.69	CAD 1,229.74	CAD 1,409.87	CAD 1,650.02	CAD 1,717.27	CAD 1,721.59
TELEFONICA EMPRESAS	CAD 913,403.28	CAD 610,457.09	CAD 558,276.52	CAD 524,567.28	CAD 574,775.09	CAD 546,589.50	CAD 608,925.61	CAD 563,322.76	CAD 538,052.17	CAD 508,510.98	CAD 507,109.83	CAD 521,868.22
VIVO EMPRESAS	CAD 34,465.00	CAD 36,331.13	CAD 39,273.61	CAD 38,294.42	CAD 40,022.29	CAD 40,994.40	CAD 33,233.46	CAD 41,480.82	CAD 32,883.08	CAD 40,788.37	CAD 39,935.45	CAD 40,464.87
Total	CAD 1,170,785.83	CAD 890,167.53	CAD 902,755.99	CAD 814,196.69	CAD 909,859.56	CAD 917,585.87	CAD 943,483.05	CAD 972,490.49	CAD 839,581.37	CAD 893,746.70	CAD 894,336.18	CAD 879,030.38

Note that this organization defined the baseline on a service provider basis, but that isn't always the case; you may define a baseline report by site, for example. The point here is that you have to start knowing your cost on a monthly basis and tracking the causes of any variation (up or down) in the costs.

The causes of the variation should be discussed in formal telecom committee meetings, where the telecom team and users discuss how the telecom usage and cost behaved in the previous month.

The consolidated telecom costs should be a by-product of controlling the bills; we should work to guarantee that the effort to have overall cost and segmented views is reasonable and the time necessary to produce the information small. The ideal situation is that the totals of the monthly telecom expenditures should be available until day ten of the subsequent month, and the committee meeting should occur by the end of the subsequent month.

Each month, the people responsible for managing telecom costs should demonstrate why the costs went up or down; they should assess the causes of this variation by type like the bellow example:

- increases due to new links contracted
- reductions due to links and voice trunks canceled
- increases due to change in capacity of data links
- increases due to increase in mobile traffic cost
- reductions due to decrease in long distance calls
- reductions due to application of new tariffs to local calls

Each item should be followed by the cost (or savings) it generated.

Besides having a report identifying the total telecom expenditures by month, we must be able to break down this cost (shown in the baseline report) by several criteria, such as by contract, by location, by resource, by invoice, or by cost center. This is important, because it enables us to trace variations in costs spotted in the baseline report. In this way, we can identify exactly what caused the increase or decrease of the cost.

Example of a cost per contract report:

Service provider	Contract description	Jan-13	Feb-13	Mar-13	Jun-13	Jul-13
BRASILSITE	ACESSO DEDICADO INTERNET - CD TUCUNARE	EUR 1,088.37	EUR 0.00	EUR 1,554.81	EUR 310.96	EUR 777.41
VIVO EMPRESAS	ACESSO INTERNET VIVO 01	EUR 0.00	EUR 0.00	EUR 0.00	EUR 0.00	EUR 0.00
VIVO EMPRESAS	ACESSO INTERNET VIVO 02	EUR 0.00	EUR 0.00	EUR 0.00	EUR 0.00	EUR 0.00
EMBRATEL	CALL CENTER 0800	EUR 14,065.27	EUR 14,391.91	EUR 13,665.42	EUR 16,591.78	EUR 16,048.46
TELEFONICA EMPRESAS	CALL CENTER 0800 TELEFONICA	EUR 357.73	EUR 278.88	EUR 258.33	EUR 236.48	EUR 262.54
EMBRATEL	CALL CENTER NUM UNICO	EUR 1,044.26	EUR 0.00	EUR 877.17	EUR 1,092.16	EUR 1,192.17
VIVO EMPRESAS	CELULARES VIVO DDD 21	EUR 109.14	EUR 116.63	EUR 104.51	EUR 141.97	EUR 207.96
VIVO EMPRESAS	CELULARES VIVO DDD 31	EUR 45.58	EUR 34.02	EUR 34.02	EUR 34.02	EUR 34.02
VIVO EMPRESAS	CELULARES VIVO DDD 41	EUR 540.52	EUR 539.51	EUR 565.36	EUR 726.44	EUR 729.22
VIVO EMPRESAS	CELULARES VIVO DDD 67	EUR 420.08	EUR 392.39	EUR 319.78	EUR 939.92	EUR 948.06
CTBC-ALGAR	CENTRAL DE RELACIONAMENTO- POP DF	EUR 8.67	EUR 8.67	EUR 8.67	EUR 8.67	EUR 8.67
CTBC-ALGAR	CENTRAL DE RELACIONAMENTO- POP GO	EUR 36.76	EUR 36.76	EUR 36.76	EUR 36.76	EUR 36.76
CTBC-ALGAR	CENTRAL DE RELACIONAMENTO- POP MG	EUR 303.55	EUR 303.80	EUR 303.54	EUR 303.55	EUR 303.62
CTBC-ALGAR	CENTRAL DE RELACIONAMENTO- POP MS	EUR 9.19	EUR 9.19	EUR 9.19	EUR 9.19	EUR 9.19
CTBC-ALGAR	CENTRAL DE RELACIONAMENTO- POP PR	EUR 9.19	EUR 9.19	EUR 9.19	EUR 9.19	EUR 9.19
CTBC-ALGAR	CENTRAL DE RELACIONAMENTO- POP RJ	EUR 32,287.29	EUR 21,783.00	EUR 26,158.14	EUR 21,866.28	EUR 20,137.25
VIVO EMPRESAS	CENTRAL DE RELACIONAMENTO-NVM	EUR 21,266.66	EUR 20,127.83	EUR 17,665.65	EUR 21,467.98	EUR 23,973.74
TELEFONICA EMPRESAS	CONTRATO VIVO CELULARES	EUR 47,125.87	EUR 47,125.87	EUR 47,125.87	EUR 47,125.87	EUR 44,522.63
OI BRT TNL	DADOS TELEFONICA 9770006	EUR 812.56	EUR 812.56	EUR 812.56	EUR 791.16	EUR 768.22
TELEFONICA EMPRESAS	DATA LINK ORBITAL	EUR 241,673.24	EUR 266,112.96	EUR 248,054.14	EUR 267,309.03	EUR 267,276.88
TELEFONICA EMPRESAS	LINKS DE DADOS - TELEFONICA	EUR 2,176.74	EUR 8,940.17	EUR 2,176.74	EUR 2,176.74	EUR 2,176.74
GIGACOM DO BRASIL LTDA.	LOCAÇÃO DE RADIOS-GDB1003/10	EUR 2,737.82	EUR 2,748.23	EUR 2,957.40	EUR 3,158.91	EUR 3,002.84
VIVO EMPRESAS	PLACAS 3G CORPORATIVAS	EUR 791.16	EUR 791.16	EUR 791.16	EUR 791.16	EUR 791.16
EMBRATEL	PRIMELINK - ORBITAL	EUR 508.24	EUR 587.97	EUR 508.55	EUR 576.43	EUR 616.37
NEXTEL	RADIO - NEXTEL	EUR 0.00	EUR 0.00	EUR 0.00	EUR 0.00	EUR 0.00
OI BRT TNL	VOZ CONSELHO RIO	EUR 155,270.88	EUR 0.00	EUR 0.00	EUR 140,465.36	EUR 9,669.67
EMBRATEL	VOZ CONTACT CENTER E LOJAS	EUR 644.11	EUR 579.92	EUR 473.85	EUR 209.13	EUR 487.18
CTBC-ALGAR	VOZ CTBC LOJAS DE SAO PAULO	EUR 369.19	EUR 232.79	EUR 204.28	EUR 208.85	EUR 196.57
CTBC-ALGAR	VOZ CTBC LOJAS GOIAS	EUR 350.38	EUR 286.74	EUR 191.08	EUR 265.80	EUR 245.78
CTBC-ALGAR	VOZ CTBC LOJAS MATO GROSSO DO SUL	EUR 1,234.06	EUR 995.95	EUR 975.36	EUR 1,002.66	EUR 1,130.86
CTBC-ALGAR	VOZ CTBC LOJAS MINAS GERAIS	EUR 0.00	EUR 229.82	EUR 264.43	EUR 644.79	EUR 665.32
TELEFONICA EMPRESAS	VOZ LOJA GOIANIA E1	EUR 567.35	EUR 748.86	EUR 679.30	EUR 801.04	EUR 805.79
SERCOMTEL	VOZ LOJA LONDRINA 01	EUR 0.00	EUR 596.66	EUR 649.90	EUR 945.41	EUR 1,709.05
OI BRT TNL	VOZ PEFISA 9700185	EUR 0.00	EUR 472.69	EUR 540.46	EUR 578.02	EUR 636.57
OI BRT TNL	VOZ REGIAO BRT MATO GROSSO	EUR 3,549.72	EUR 3,435.58	EUR 3,627.66	EUR 3,468.23	EUR 3,343.60
OI BRT TNL	VOZ REGIAO BRT MATO GROSSO DO SUL	EUR 10,299.04	EUR 10,113.09	EUR 9,657.86	EUR 9,281.55	EUR 11,053.00
OI BRT TNL	VOZ REGIAO I OI MINAS GERAIS	EUR 991.05	EUR 1,017.53	EUR 1,156.77	EUR 1,060.48	EUR 1,101.65
OI BRT TNL	VOZ REGIAO II BRT GOIÁS	EUR 13,106.13	EUR 12,479.50	EUR 12,724.29	EUR 11,405.10	EUR 11,533.58
OI BRT TNL	VOZ REGIAO II BRT PARANÁ	EUR 3,483.75	EUR 2,885.52	EUR 2,749.80	EUR 2,334.58	EUR 2,489.36
OI BRT TNL	VOZ REGIAO II BRT SANTA CATARINA	EUR 63,863.05	EUR 51,511.89	EUR 59,976.06	EUR 52,477.24	EUR 50,889.67
TELEFONICA EMPRESAS	VOZ TELEFONICA 9770006	EUR 0.00	EUR 0.00	EUR 0.00	EUR 0.00	EUR 0.00
CTBC-ALGAR	VOZ TELEMARKETING	EUR 0.00	EUR 0.00	EUR 0.00	EUR 0.00	EUR 0.00
Total		EUR 621,481.68	EUR 470,746.43	EUR 457,877.25	EUR 610,862.08	EUR 477,623.20

Example of a cost per resource report:

Telco ID	Last mile description	Jan-13	Feb-13	Mar-13	Apr-13	May-13	Jun-13
8007022066	TRONCO DIGITAL E1	R$ 1.74	R$ 0.00	R$ 0.00	R$ 2.66	R$ 3.31	R$ 12.85
8007029248	TRONCO DIGITAL E1	R$ 859.19	R$ 743.13	R$ 694.81	R$ 740.18	R$ 749.47	R$ 763.72
8007240066	TRONCO DIGITAL E1	R$ 5,953.19	R$ 6,004.78	R$ 4,838.25	R$ 6,119.27	R$ 6,345.31	R$ 6,331.03
8007249200	TRONCO DIGITAL E1	R$ 38,186.93	R$ 40,048.10	R$ 38,143.75	R$ 46,336.11	R$ 44,343.76	R$ 44,721.54
8007254005	TRONCO DIGITAL E1	R$ 230.36	R$ 228.95	R$ 268.74	R$ 158.01	R$ 167.15	R$ 167.84
8009316880	TRONCO DIGITAL E1	R$ 1,150.40	R$ 896.82	R$ 830.75	R$ 760.47	R$ 844.29	R$ 803.75
110000071250997	PACOTE SMART	R$ 1,938.54	R$ 209.60	R$ 209.59	R$ 209.59	R$ 209.59	R$ 209.59
110000071313999	PACOTE SMART	R$ 0.00	R$ 276.03	R$ 276.02	R$ 276.02	R$ 276.03	R$ 276.02
110000071314190	PACOTE SMART	R$ 0.00	R$ 276.03	R$ 276.02	R$ 276.02	R$ 276.03	R$ 276.02
110000071468393	PACOTE SMART	R$ 2,687.28	R$ 276.03	R$ 276.02	R$ 276.02	R$ 276.03	R$ 276.02
110000071469194	PACOTE SMART	R$ 1,938.54	R$ 209.60	R$ 209.59	R$ 209.59	R$ 209.59	R$ 209.59
110000071600290	PACOTE SMART	R$ 0.00	R$ 0.00	R$ 0.00	R$ 0.00	R$ 0.00	R$ 0.00
110000071635698	PACOTE SMART	R$ 2,687.29	R$ 276.03	R$ 276.04	R$ 276.04	R$ 276.03	R$ 276.03
110000071648794	LINK FRAME RELAY	R$ 1,938.54	R$ 1,215.02	R$ 1,215.00	R$ 1,215.01	R$ 1,215.01	R$ 1,215.00
110000071720298	PACOTE SMART	R$ 0.00	R$ 0.00	R$ 0.00	R$ 0.00	R$ 0.00	R$ 0.00
110000071720592	PACOTE SMART	R$ 1,938.54	R$ 209.60	R$ 209.59	R$ 209.59	R$ 209.59	R$ 209.59
110000071728492	PACOTE SMART	R$ 0.00	R$ 0.00	R$ 0.00	R$ 0.00	R$ 0.00	R$ 0.00
110000071906496	PACOTE SMART	R$ 1,938.54	R$ 209.60	R$ 209.59	R$ 209.59	R$ 209.59	R$ 209.59
110000071906692	PACOTE SMART	R$ 1,938.54	R$ 209.60	R$ 209.59	R$ 209.59	R$ 209.59	R$ 209.59
110000071936790	PACOTE SMART	R$ 0.00	R$ 209.60	R$ 209.59	R$ 209.59	R$ 209.59	R$ 209.59
110000074597699	PACOTE SMART	R$ 193.85	R$ 209.60	R$ 209.59	R$ 209.59	R$ 209.59	R$ 209.59
110000076513295	PACOTE SMART	R$ 0.00	R$ 0.00	R$ 0.00	R$ 0.00	R$ 0.00	R$ 0.00
110000076805496	PACOTE SMART	R$ 0.00	R$ 209.60	R$ 209.59	R$ 209.59	R$ 209.59	R$ 209.59
110000076807490	PACOTE SMART	R$ 0.00	R$ 209.60	R$ 209.59	R$ 209.59	R$ 209.59	R$ 209.59
110000076807599	PACOTE SMART	R$ 0.00	R$ 209.60	R$ 209.59	R$ 209.59	R$ 209.59	R$ 209.59
110000076807893	PACOTE SMART	R$ 0.00	R$ 209.60	R$ 209.59	R$ 209.59	R$ 209.59	R$ 209.59
110000076808095	PACOTE SMART	R$ 0.00	R$ 209.60	R$ 209.59	R$ 209.59	R$ 209.59	R$ 209.59
110000076808291	PACOTE SMART	R$ 0.00	R$ 209.60	R$ 209.59	R$ 209.59	R$ 209.59	R$ 209.59
110000076809790	PACOTE SMART	R$ 0.00	R$ 209.60	R$ 209.59	R$ 209.59	R$ 209.59	R$ 209.59
110000076810093	PACOTE SMART	R$ 0.00	R$ 209.60	R$ 209.59	R$ 209.59	R$ 209.59	R$ 209.59
110000076811395	PACOTE SMART	R$ 0.00	R$ 209.60	R$ 209.59	R$ 209.59	R$ 209.59	R$ 209.59
110000076812196	PACOTE SMART	R$ 0.00	R$ 209.60	R$ 209.59	R$ 209.59	R$ 209.59	R$ 209.59
110000076814898	PACOTE SMART	R$ 0.00	R$ 209.60	R$ 209.59	R$ 209.59	R$ 209.59	R$ 209.59
Total		R$ 63,581.47	R$ 53,804.12	R$ 50,868.02	R$ 60,208.43	R$ 58,545.04	R$ 58,892.44

Example of a cost per cost center report:

Cost center code	Cost center name	Jan-13	Feb-13	Mar-13	Apr/13	May/13	Jul-13
10311	DIRETORIA SUPERINTENDENCIA	R$ 3,311.17	R$ 3,165.59	R$ 2,493.67	R$ 2,375.25	R$ 2,529.92	R$ 2,503.96
10321	DIRETORIA COMERCIAL	R$ 2,515.95	R$ 2,396.40	R$ 1,809.16	R$ 3,844.59	R$ 2,046.61	R$ 1,978.82
10331	DIRETORIA OPERACOES	R$ 2,316.75	R$ 3,021.86	R$ 2,247.12	R$ 2,259.53	R$ 2,321.63	R$ 2,842.88
10341	DIRETORIA FINANCEIRA	R$ 1,008.51	R$ 1,361.83	R$ 980.36	R$ 1,999.88	R$ 960.58	R$ 997.76
10351	DIRETORIA CONTROLADORIA	R$ 934.67	R$ 918.17	R$ 920.79	R$ 1,825.60	R$ 827.55	R$ 1,750.63
10401	GESTAO DE PROJETOS	R$ 2,586.51	R$ 2,172.04	R$ 2,107.52	R$ 2,110.11	R$ 2,096.52	R$ 1,512.21
107-14813	CD GOLDEN	R$ 10,166.41	R$ 0.00	R$ 14,523.44	R$ 2,904.69	R$ 0.00	R$ 0.00
10701	SEGUROS	R$ 770.85	R$ 839.75	R$ 767.68	R$ 997.41	R$ 845.93	R$ 753.68
11101	COMPRAS LAR	R$ 3,302.16	R$ 3,447.20	R$ 3,268.03	R$ 3,126.84	R$ 2,799.19	R$ 2,765.67
11501	COMPRAS VESTUARIO	R$ 2,992.79	R$ 4,789.19	R$ 4,599.05	R$ 7,783.78	R$ 6,459.68	R$ 6,604.94
11701	COMPRAS ELETRO	R$ 6,418.75	R$ 6,294.95	R$ 6,191.34	R$ 6,200.72	R$ 6,213.18	R$ 6,365.27
119-14801	CD TUCUNARÉ	R$ 0.00	R$ 0.00	R$ 0.00	R$ 29,521.30	R$ 24,399.37	R$ 24,399.37
12201	IMPORTACAO	R$ 1,747.78	R$ 1,240.13	R$ 1,432.85	R$ 1,978.79	R$ 1,847.67	R$ 3,692.24
12501	DESENVOLVIMENTO DE PRODUTOS	R$ 0.00	R$ 0.00	R$ 0.00	R$ 0.00	R$ 0.00	R$ 0.00
12701	MARKETING	R$ 8,633.57	R$ 9,134.34	R$ 8,751.33	R$ 9,445.20	R$ 9,170.01	R$ 10,077.75
12801	VISUAL E MERCHANDISING	R$ 20,260.34	R$ 17,272.49	R$ 12,380.30	R$ 18,749.32	R$ 13,609.36	R$ 15,378.34
12901	DESENVOLVIMENTO DE VENDAS	R$ 40,947.55	R$ 22,894.40	R$ 19,192.75	R$ 35,022.40	R$ 36,300.78	R$ 45,364.48
13101	PLANEJAMENTO COMERCIAL	R$ 2,450.98	R$ 2,450.98	R$ 2,431.80	R$ 2,447.40	R$ 2,413.94	R$ 2,671.47
13801	ENGENHARIA	R$ 31,829.74	R$ 25,792.69	R$ 23,135.40	R$ 27,035.12	R$ 30,458.52	R$ 28,643.24
15101	TECNOLOGIA	R$ 793,608.12	R$ 826,441.66	R$ 813,284.91	R$ 817,145.18	R$ 1,081,750.57	R$ 801,412.99
15201	UCOLL	R$ 65,216.24	R$ 64,152.92	R$ 68,152.82	R$ 67,641.98	R$ 67,678.72	R$ 66,512.23
15301	RECURSOS HUMANOS	R$ 7,353.01	R$ 9,002.29	R$ 8,475.21	R$ 9,019.90	R$ 8,250.18	R$ 8,138.50
15501	PATRIMONIO	R$ 5,719.56	R$ 6,005.27	R$ 5,716.83	R$ 5,800.54	R$ 5,797.09	R$ 6,240.05
15601	CENTRAL CLIENTE	R$ 117,446.62	R$ 121,755.78	R$ 116,055.62	R$ 139,444.54	R$ 134,204.54	R$ 135,272.65
15701	COMPRAS SUPRIMENTOS E SERVICOS	R$ 1,211.89	R$ 1,098.06	R$ 1,060.21	R$ 1,063.64	R$ 1,155.66	R$ 1,119.18
16101	AUDITORIA INTERNA	R$ 1,093.50	R$ 1,133.41	R$ 1,197.69	R$ 1,132.65	R$ 1,170.47	R$ 1,141.16
16301	FINANCAS	R$ 2,216.31	R$ 1,906.75	R$ 3,002.63	R$ 2,823.30	R$ 1,392.74	R$ 4,468.57
16401	PRODUTOS FINANCEIROS	R$ 97,261.42	R$ 104,005.99	R$ 103,748.60	R$ 104,048.22	R$ 103,804.55	R$ 112,463.13
16501	CREDITO E COBRANCA	R$ 3,325.26	R$ 3,490.59	R$ 3,362.50	R$ 3,523.53	R$ 3,821.64	R$ 5,059.94
16601	CENTRAL DE RELACIONAMENTO	R$ 1,718,656.48	R$ 1,349,991.01	R$ 1,641,775.30	R$ 1,485,649.52	R$ 1,695,010.09	R$ 1,615,988.56
16701	INFORMACOES GERENCIAIS	R$ 624.39	R$ 615.10	R$ 617.74	R$ 616.98	R$ 602.37	R$ 612.16
16704	INFORMAÇÕES GERÊNCIAS 3	R$ 796.41	R$ 818.13	R$ 820.78	R$ 821.33	R$ 788.42	R$ 818.05
16706	INFORMAÇÕES GERÊNCIAS 2	R$ 1,036.01	R$ 1,108.78	R$ 1,159.20	R$ 1,124.20	R$ 1,083.36	R$ 1,222.84
16901	CONTABILIDADE	R$ 902.49	R$ 999.01	R$ 842.94	R$ 865.54	R$ 832.11	R$ 883.34
17101	JURIDICO	R$ 1,804.13	R$ 1,414.18	R$ 2,133.75	R$ 1,101.81	R$ 1,164.84	R$ 1,224.67
17301	TRIBUTARIO	R$ 1,854.64	R$ 1,783.27	R$ 1,570.56	R$ 1,746.27	R$ 1,512.50	R$ 1,344.00
18601	COBRANCA	R$ 203.04	R$ 203.04	R$ 203.04	R$ 203.04	R$ 203.04	R$ 203.04
18801	CD CADIRIRI	R$ 2,002.23	R$ 1,466.29	R$ 1,912.36	R$ 1,732.12	R$ 1,735.52	R$ 1,499.63
19201	LOJAS	R$ 2,729,110.75	R$ 2,896,459.44	R$ 2,760,803.66	R$ 2,769,020.42	R$ 2,837,007.36	R$ 2,776,856.28
RATEIO	RATEIO ADM BELA CINTRA	R$ 766.26	R$ 626.48	R$ 514.57	R$ 337.06	R$ 0.00	R$ 0.00
RATEIO CONSELHO	50%10143 E 50%10163	R$ 5,794.85	R$ 6,265.56	R$ 6,265.50	R$ 6,265.50	R$ 6,265.50	R$ 6,265.50
RATEIO CONSELHO RJ	50%10153 E 50%10143	R$ 0.00	R$ 0.00	R$ 0.00	R$ 0.00	R$ 0.00	R$ 0.00
Total		R$ 5,700,198.09	R$ 5,507,935.02	R$ 5,649,909.01	R$ 5,580,755.20	R$ 6,100,531.71	R$ 5,707,490.81

15.4.2   Volume and Profile of the Traffic

Another important set of reports is the traffic usage. These reports usually refer to voice traffic and identify the quantity, duration, and costs per type of call per month. We usually divide these reports by service, such as call center operations, administrative, and so forth. These reports should allow us to follow the volumes month by month, spotting any change in the traffic pattern. Usually the traffic reports are generated based on two sources:

- the billing system
- the telco bills logs

Here it is important to normalize the cycle of these sources in order to avoid discrepancies. Therefore, if we want to check the number of calls per type in the call center operation, we should generate the report from the billing system using the same period of the bills. Doing that makes it possible to compare the cost indicated by the billing system with the values charged by the providers. These reports usually encompass three main variables: quantity of calls, duration in minutes, and cost.

Usually we divide the type of calls in the following manner:

Inbound

- from mobile local
- from mobile long distance
- from a fixed line local
- from a fixed line long distance

Outbound

- to local mobile phones
- to long distance mobile phones
- to local fixed lines
- to long distance fixed lines

15.4.3   The Analysis of the Reports

Although most companies have the tools to generate several types of reports, it is advisable to have defined sets of reports generated in a defined schedule. This is important because it normalizes the information and sets a routine of generating, forwarding, analyzing, and discussing those reports (via telecom committee meetings). We should have a predefined schedule of generating the reports, and in case we need to drill down some information, we can generate off-the-schedule additional reports. Most cost reports are generated monthly and should be generated after all the monthly invoices have been processed. There are typically four groups of telecom cost reduction strategies:

- pressure the service providers and hardware vendors, trying to guarantee low prices (for example, bargaining and negotiating hard)
- enhance the internal control over usage of the services available (for example, billing systems)
- increase the control over the service providers to make sure that the organization is only paying for what was really used at the agreed value for each service (for example, telephone bills auditing)
- carefully manage traffic and network design, emphasizing aspects such as standardization and simplification of the technical environments (for example, least cost routing)

Usually we have at least one of these strategies always being implemented in a given time in a large organization, and therefore it is important to link the telecom expense management process with the cost reduction strategies being implemented.

This is crucial because each action taken (contract negotiations, for example) must be followed in order to certify that the gains foreseen are really happening. This is a key point; although telecom cost management aims to spot undue and untypical charges, the main objective should be to check if the strategies being deployed by the organization to reduce usage or lower costs are really yielding the expected results.

However, it is important to keep in mind that before any cost reduction strategy can be followed, it is necessary to identify the current telecom costs (the reference or baseline). Knowing how much is spent today is the only way to confirm whether the actions taken generate the planned gains.

In addition, the knowledge of the current expenditure and volume is the basis over which all decisions are made. Only knowing how much you spend in total and with what services will you be able to identify what can be done to reduce the current costs.

In other words, once you know your current costs and start implementing cost-reduction strategies, following the costs is the only way to know if what you are doing is yielding the expected results.

This statement may sound straightforward; however, what is very often seen is telecom expense management teams completely locked into processing, registering, and looking for mistakes in the bills. This mind-set makes them completely lose track of the cost-reduction strategies being implemented by the organization. In addition, this mind-set jeopardizes any effort to think about possible initiatives to reduce telecom costs. These situations are what we could call sterile or bureaucratic control: control that doesn't support the real needs of the organization.

Knowing the monthly cost of the telecom infrastructure is pretty obvious, but many organizations don't have a clear view about their telecom cost. This happens for several reasons:

- Telecom is managed in a decentralized way (business units can contract and pay for their own resources, and telecom costs are inappropriately associated with other cost centers).
- There isn't a centralized inventory of the resources and telecom contracts.
- There isn't any traffic management.

15.4.4 How to Structure the Information

When analyzing the costs, it is important to understand clearly the components being charged. Typically we can classify all charges into one of the following six categories:

1. Recurrent values
2. Variable values dependent on usage
3. Eventual fees
4. Taxes
5. Penalties, credits and Discounts

Recurrent values: Those are the values associated with the subscription of services, which have the same value every month (or are supposed to). In this category, we have the charges associated with data links, voice trunk subscriptions, and equipment rental or maintenance.

Variable values (usage dependent): Those are the values associated with traffic usage; these values usually vary each month (they depend on the usage), although within a typical range. Although there are variations, these values tend to stay within a recognizable range, and following the monthly expenditures allows you to identify discrepancies between what was charged and what would be typical. Some organizations audit the bills only when discrepancies are spotted (values or volumes).

Eventual fees: Those are the fees related to one-time services like installation or canceling of services.

Taxes: In some bills, the taxes are indicated separately. This may be important information, given the fact that in some countries, there are special tax rebates or tax breaks for call center operations.

Penalties, credits and discounts: Those are charges or credits usually associated with late payment or no compliance with SLAs.

Here it is important to separate the charges into two types in each category: how much was charged and how much is actually due. It is an important distinction, which must be registered. At this point, auditing comes to play.

Another important aspect to discuss is the cycle of the bills. When identifying the monthly telecom expenditures, a very common misleading factor is the fact that telco invoices encompass charges of different periods of time. Therefore, when analyzing volumes, it is important to understand exactly the period when the calls were placed and when they were charged. When comparing volumes and cost, it is important to make sure that the cycles are correct.

We should make an effort to put all invoices within the same cycle. We often have situations where the cycles of several contracts are not the same. That means each month we will be paying for calls and services that actually took place in a different time span (often with some overlap). This fact may generate distortions when comparing monthly volumes and costs or when comparing bills with billing systems logs.

For example, during a given month, for some reason, we forwarded more calls to Telco A than Telco B (as usual). If Telco A has a charging period different from Telco B, the calls made through Telco A may be charged in the subsequent month instead of the current one (as would happen if we had used Telco B). This fact generates an illusion that this month the telecom cost was lower than usual. By the same token, next month we will notice that the telecom cost was higher.

Chapter 16: Closing Words

We hope that this book proves useful as a reference guide. We tried to consolidate a large set of information in a coherent way in a relatively small space. In an endeavor like this, it is almost impossible to include everything; however, we believe we achieved our goal: to provide a useful tool through which IT and telecom managers can improve the efficiency of the management of their infrastructure.

Many of the topics discussed demand some previous understanding, and many of the opinions expressed may be somewhat controversial in their interpretation by the authors. In addition, we are very aware that organizational realities may drive a politically based decision, which takes away the Cartesian line of thought expressed in this book.

We took the perspective of telecom managers; this book was written mostly for them. We know that many of the topics covered in the book may be viewed from a different perspective, depending on which hat you are wearing. Hardware vendors and telco representatives may also benefit from this book by improving their understanding of the challenges faced by their clients.

Bibliography

Bayer, Michael. *CTI Solutions and Systems: How to Put Computer Telephony Integration to Work*. McGraw-Hill.

Brosnan, Michael, John Messina, and Ellen Block. *Telecommunications Expense Management*. Miller Freeman.

Harnett, Donald L. *Statistical Analysis for Business and Economics*. Addison-Wesley.

Network Analysis Corporation. *ARPANET: Design, Operation, Management, and Performance*. New York: 1973.

Sharma, Roshan L. *Network Topology Optimization: The Art and Science of Network Design*. VNR Computer Library, 1990.

Software ARIETE®-WANOPT®

Software TRMS® Telecommunications Resources Management System-WANOPT®

S.C. Strother, S. C. *Telecom Cost Management*. Arthech House, 2002.

Wide Area Network Methodology®-WANOPT®